Cram101 Textbook Outlines to accompany:

Life: The Science of Biology

Purves, Sadava, Orians, Heller, 6th Edition

An Academic Internet Publishers (AIPI) publication (c) 2007.

You have a discounted membership at www.Cram101.com with this book.

Get all of the practice tests for the chapters of this textbook, and access in-depth reference material for writing essays and papers. Here is an example from a Cram101 Biology text:

When you need problem solving help with math, stats, and other disciplines, www.Cram101.com will walk through the formulas and solutions step by step.

With Cram101.com online, you also have access to extensive reference material.

You will nail those essays and papers. Here is an example from a Cram101 Biology text:

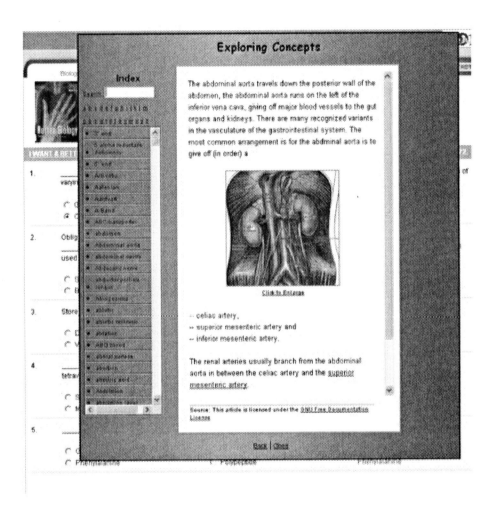

Learning System

Cram101 Textbook Outlines is a learning system. The notes in this book are the highlights of your textbook, you will never have to highlight a book again.

How to use this book. Take this book to class, it is your notebook for the lecture. The notes and highlights on the left hand side of the pages follow the outline and order of the textbook. All you have to do is follow along while your intructor presents the lecture. Circle the items emphasized in class and add other important information on the right side. With Cram101 Textbook Outlines you'll spend less time writing and more time listening. Learning becomes more efficient.

Cram101.com Online

Increase your studying efficiency by using Cram101.com's practice tests and online reference material. It is the perfect complement to Cram101 Textbook Outlines. Use self-teaching matching tests or simulate in-class testing with comprehensive multiple choice tests, or simply use Cram's true and false tests for quick review. Cram101.com even allows you to enter your in-class notes for an integrated studying format combining the textbook notes with your class notes.

Visit **www.Cram101.com**, click Sign Up at the top of the screen, and enter **DK73DW210** in the promo code box on the registration screen. Access to www.Cram101.com is normally $9.95, but because you have purchased this book, your access fee is only $4.95. Sign up and stop highlighting textbooks forever.

Life: The Science of Biology
Purves, Sadava, Orians, Heller, 6th

CONTENTS

Life: The Science of Biology
Purves, Sadava, Orians, Heller, 6th

CONTENTS (continued)

Radioactive	A term used to describe the property of releasing energy or particles from an unstable atom is called radioactive.
Evolution	In biology, evolution is the process by which novel traits arise in populations and are passed on from generation to generation. Its action over large stretches of time explains the origin of new species and ultimately the vast diversity of the biological world.
Adaptation	A biological adaptation is an anatomical structure, physiological process or behavioral trait of an organism that has evolved over a period of time by the process of natural selection such that it increases the expected long-term reproductive success of the organism.
Species	Group of similarly constructed organisms capable of interbreeding and producing fertile offspring is a species.
Climate	Weather condition of an area including especially prevailing temperature and average daily/yearly rainfall over a long period of time is called climate.
Natural selection	Natural selection is the process by which biological individuals that are endowed with favorable or deleterious traits end up reproducing more or less than other individuals that do not possess such traits.
Cyanobacteria	Cyanobacteria are a phylum of bacteria that obtain their energy through photosynthesis. They are often referred to as blue-green algae, even though it is now known that they are not directly related to any of the other algal groups, which are all eukaryotes.
Photosynthesis	Photosynthesis is a biochemical process in which plants, algae, and some bacteria harness the energy of light to produce food. Ultimately, nearly all living things depend on energy produced from photosynthesis for their nourishment, making it vital to life on Earth.
Unicellular	Unicellular organisms carry out all the functions of life. Unicellular species are those whose members consist of a single cell throughout their life cycle. This latter qualification is significant since most multicellular organisms consist of a single cell at the beginning of their life cycles.
Reproduction	Biological reproduction is the biological process by which new individual organisms are produced. Reproduction is a fundamental feature of all known life; each individual organism exists as the result of reproduction by an antecedent.
Bacteria	The domain that contains procaryotic cells with primarily diacyl glycerol diesters in their membranes and with bacterial rRNA. Bacteria also is a general term for organisms that are composed of procaryotic cells and are not multicellular.
Eukaryotic cell	Eukaryotic cell refers to a type of cell that has a membrane-enclosed nucleus and other membrane enclosed organelles. All organisms except bacteria and archaea are composed of eukaryotic cells.
Cell	The cell is the structural and functional unit of all living organisms, and is sometimes called the "building block of life."
Element	A chemical element, often called simply element, is a chemical substance that cannot be divided or changed into other chemical substances by any ordinary chemical technique. An element is a class of substances that contain the same number of protons in all its atoms.
Meiosis	In biology, meiosis is the process that transforms one diploid cell into four haploid cells in eukaryotes in order to redistribute the diploid's cell's genome. Meiosis forms the basis of sexual reproduction and can only occur in eukaryotes.
Multicellular	Multicellular organisms are those organisms consisting of more than one cell, and having differentiated cells that perform specialized functions. Most life that can be seen with the naked eye is multicellular, as are all animals (i.e. members of the kingdom Animalia) and plants (i.e. members of the kingdom Plantae).
Gene	Gene refers to a discrete unit of hereditary information consisting of a specific nucleotide sequence

	in DNA . Most of the genes of a eukaryote are located in its chromosomal DNA; a few are carried by the DNA of mitochondria and chloroplasts.
Control processes	Mechanisms that ensure an organism will carry out all life activities in the proper sequence and at the proper rate are called control processes.
Population	Group of organisms of the same species occupying a certain area and sharing a common gene pool is referred to as population.
Nerve	A nerve is an enclosed, cable-like bundle of nerve fibers or axons, which includes the glia that ensheath the axons in myelin.
Fossil	A preserved remnant or impression of an organism that lived in the past is referred to as fossil.
Archaea	The Archaea are a major division of living organisms. Although there is still uncertainty in the exact phylogeny of the groups, Archaea, Eukaryotes and Bacteria are the fundamental classifications in what is called the three-domain system.
Domain	In biology, a domain is the top-level grouping of organisms in scientific classification.
Genus	In biology, a genus is a taxonomic grouping. That is, in the classification of living organisms, a genus is considered to be distinct from other such genera. A genus has one or more species: if it has more than one species these are likely to be morphologically more similar than species belonging to different genera.
Scientific name	The name of an organism formed from the two smallest major taxonomic categories-the genus and the species is a scientific name.
Biology	Biology is the branch of science dealing with the study of life. It is concerned with the characteristics, classification, and behaviors of organisms, how species come into existence, and the interactions they have with each other and with the environment.
Genetic code	The genetic code is a set of rules that maps DNA sequences to proteins in the living cell, and is employed in the process of protein synthesis. Nearly all living things use the same genetic code, called the standard genetic code, although a few organisms use minor variations of the standard code.
Extinction	In biology and ecology, extinction is the ceasing of existence of a species or group of taxa. The moment of extinction is generally considered to be the death of the last individual of that species.The death of all members of a species is extinction.

Atom	An atom is the smallest possible particle of a chemical element that retains its chemical properties.
Nucleus	In cell biology, the nucleus is found in all eukaryotic cells that contains most of the cell's genetic material. The nucleus has two primary functions: to control chemical reactions within the cytoplasm and to store information needed for cellular division.
Proton	Positive subatomic particle, located in the nucleus and having a weight of approximately one atomic mass unit is referred to as a proton.
Electron	The electron is a light fundamental subatomic particle that carries a negative electric charge. The electron is a spin-1/2 lepton, does not participate in strong interactions and has no substructure.
Neutron	An electrically neutral particle , found in the nucleus of an atom is referred to as neutron.
Element	A chemical element, often called simply element, is a chemical substance that cannot be divided or changed into other chemical substances by any ordinary chemical technique. An element is a class of substances that contain the same number of protons in all its atoms.
Nerve	A nerve is an enclosed, cable-like bundle of nerve fibers or axons, which includes the glia that ensheath the axons in myelin.
Mass number	The mass number is the number of protons plus the number of neutrons in an atomic nucleus. The mass number is unique for each isotope of an element and is written either after the element name or as a superscript to the left of an element's symbol.
Atomic number	In chemistry and physics, the atomic number (Z) is the number of protons found in the nucleus of an atom. In an atom of neutral charge, the number of electrons also equals the atomic number.
Isotope	An isotope is a form of an element whose nuclei have the same atomic number - the number of protons in the nucleus - but different mass numbers because they contain different numbers of neutrons.
Radiation	The emission of electromagnetic waves by all objects warmer than absolute zero is referred to as radiation.
Electron shell	The electron shell is a group of atomic orbitals with the same value of the principal quantum number n. The electron shell determines the chemical properties of the atom.
Molecule	A molecule is the smallest particle of a pure chemical substance that still retains its chemical composition and properties.
Chemical bond	Chemical bond refers to an attraction between two atoms resulting from a sharing of outer-shell electrons or the presence of opposite charges on the atoms. The bonded atoms gain complete outer electron shells.
Hydrogen bond	A hydrogen bond is a type of attractive intermolecular force that exists between two partial electric charges of opposite polarity. Although stronger than most other intermolecular forces, the typical hydrogen bond is much weaker than both the ionic bond and the covalent bond.
Hydrogen	Hydrogen is a chemical element in the periodic table that has the symbol H and atomic number 1. At standard temperature and pressure it is a colorless, odorless, nonmetallic, univalent, tasteless, highly flammable diatomic gas.
Ion	Ion refers to an atom or molecule that has gained or lost one or more electrons, thus acquiring an electrical charge.
Polar molecule	Molecule that displays an uneven distribution of electrons over its structure, for example,

	water is a polar molecule.
Hydrocarbon	A chemical compound composed only of the elements carbon and hydrogen is called hydrocarbon.
Chemical reaction	Chemical reaction refers to a process leading to chemical changes in matter; involves the making and/or breaking of chemical bonds.
Calorie	Calorie refers to the amount of energy that raises the temperature of 1 g of water by 1°C.
Oxidation	Oxidation refers to the loss of electrons from a substance involved in a redox reaction; always accompanies reduction.
Solvent	A solvent is a liquid that dissolves a solid, liquid, or gaseous solute, resulting in a solution. The most common solvent in everyday life is water.
Mole	The atomic weight of a substance, expressed in grams. One mole is defined as the mass of 6.02223×10^{23} atoms.
Specific heat	The amount of energy required to raise the temperature of 1 gram of a substance by 1 °C is specific heat.
Leaf	In botany, a leaf is an above-ground plant organ specialized for photosynthesis. For this purpose, a leaf is typically flat (laminar) and thin, to expose the chloroplast containing cells (chlorenchyma tissue) to light over a broad area, and to allow light to penetrate fully into the tissues.
Surface tension	A measure of how difficult it is to stretch or break the surface of a liquid is referred to as surface tension.
Biochemistry	Biochemistry studies how complex chemical reactions give rise to life. It is a hybrid branch of chemistry which specialises in the chemical processes in living organisms.
Acid	An acid is a water-soluble, sour-tasting chemical compound that when dissolved in water, gives a solution with a pH of less than 7.
Bicarbonate ion	The bicarbonate ion consists of one central carbon atom surrounded by three identical oxygen atoms in a trigonal planar arrangement, with a hydrogen atom attached to one of the oxygens.
Reactant	A reactant is any substance initially present in a chemical reaction. These reactants react with each other to form the products of a chemical reaction. In a chemical equation, the reactants are the elements or compounds on the left hand side of the reaction equation.
Buffer	A chemical substance that resists changes in pH by accepting H^+ ions from or donating H^+ ions to solutions is called a buffer.
Isomer	An isomer is a molecule with the same chemical formula and often with the same kinds of bonds between atoms, but in which the atoms are arranged differently. That is to say, they have different structural formula.
Amino acid	An amino acid is any molecule that contains both amino and carboxylic acid functional groups. They are the basic structural building units of proteins. They form short polymer chains called peptides or polypeptides which in turn form structures called proteins.

Molecule	A molecule is the smallest particle of a pure chemical substance that still retains its chemical composition and properties.
Lipid	Lipid is one class of aliphatic hydrocarbon-containing organic compounds essential for the structure and function of living cells. They are characterized by being water-insoluble but soluble in nonpolar organic solvents.
Carbohydrate	Carbohydrate is a chemical compound that contains oxygen, hydrogen, and carbon atoms. They consist of monosaccharide sugars of varying chain lengths and that have the general chemical formula $C_n(H_2O)_n$ or are derivatives of such.
Nucleic acid	A nucleic acid is a complex, high-molecular-weight biochemical macromolecule composed of nucleotide chains that convey genetic information. The most common are deoxyribonucleic acid (DNA) and ribonucleic acid (RNA). They are found in all living cells and viruses.
Acid	An acid is a water-soluble, sour-tasting chemical compound that when dissolved in water, gives a solution with a pH of less than 7.
Macromolecule	A macromolecule is a molecule of high relative molecular mass, the structure of which essentially comprises the multiple repetition of units derived, actually or conceptually, from molecules of low relative molecular mass.
Functional group	Functional group is a submolecular structural motif, characterized by specific elemental composition and connectivity, that confer reactivity upon the molecule that contains it.
Protein	A protein is a complex, high-molecular-weight organic compound that consists of amino acids joined by peptide bonds. They are essential to the structure and function of all living cells and viruses. Many are enzymes or subunits of enzymes.
Catalysis	Catalysis is the acceleration of the reaction rate of a chemical reaction by means of a substance, called a catalyst, that is itself not consumed by the overall reaction.
Chemical reaction	Chemical reaction refers to a process leading to chemical changes in matter; involves the making and/or breaking of chemical bonds.
Polypeptide	Polypeptide refers to polymer of many amino acids linked by peptide bonds.
Amino acid	An amino acid is any molecule that contains both amino and carboxylic acid functional groups. They are the basic structural building units of proteins. They form short polymer chains called peptides or polypeptides which in turn form structures called proteins.
Hydrophobic	Hydrophobic refers to being electrically neutral and nonpolar, and thus prefering other neutral and nonpolar solvents or molecular environments. Hydrophobic is often used interchangeably with "oily" or "lipophilic."
Covalent bond	A covalent bond is an intramolecular form of chemical bonding characterized by the sharing of one or more pairs of electrons between two elements, producing a mutual attraction that holds the resultant molecule together.
Disulfide bridge	Disulfide bridge is a strong covalent bond between two sulfhydryl groups. This bond is very important to the folding, structure, and function of proteins.
Atom	An atom is the smallest possible particle of a chemical element that retains its chemical properties.
Amino group	An amino group is an ammonia-like functional group composed of a nitrogen and two hydrogen atoms covalently linked. $-NH_2$
Carboxyl group	In an organic molecule, a functional group consisting of an oxygen atom doublebonded to a carbon atom that is also bonded to a hydroxyl group is referred to as a carboxyl group.
Carboxyl	A carboxyl is the univalent radical -COOH; present in and characteristic of organic acids.

Peptide bond	A peptide bond is a chemical bond formed between two molecules when the carboxyl group of one molecule reacts with the amino group of the other molecule, releasing a molecule of water.
Peptide	Peptide is the family of molecules formed from the linking, in a defined order, of various amino acids. The link between one amino acid residue and the next is an amide bond, and is sometimes referred to as a peptide bond.
Primary structure	The primary structure of an unbranched biopolymer, such as a molecule of DNA, RNA or protein, is the specific nucleotide or peptide sequence from the beginning to the end of the molecule.
Helix	A helix is a twisted shape like a spring, screw or a spiral staircase. They are important in biology, as DNA and many proteins have spiral substructures, known a alpha helix.
Secondary structure	Secondary structure refers to the second level of protein structure; the regular patterns of coils or folds of a polypeptide chain.
Pleated sheet	A form of secondary structure exhibited by certain proteins, such as silk, characterized by antiparallel molecules with zigzag structure is called a pleated sheet.
Hemoglobin	Hemoglobin is the iron-containing oxygen-transport metalloprotein in the red cells of the blood in mammals and other animals. Hemoglobin transports oxygen from the lungs to the rest of the body, such as to the muscles, where it releases the oxygen load.
Cell	The cell is the structural and functional unit of all living organisms, and is sometimes called the "building block of life."
Virus	Obligate intracellular parasite of living cells consisting of an outer capsid and an inner core of nucleic acid is referred to as virus. The term virus usually refers to those particles that infect eukaryotes whilst the term bacteriophage or phage is used to describe those infecting prokaryotes.
Hydroxyl group	The term hydroxyl group is used to describe the functional group -OH when it is a substituent in an organic compound.
Biochemistry	Biochemistry studies how complex chemical reactions give rise to life. It is a hybrid branch of chemistry which specialises in the chemical processes in living organisms.
Protease	Protease refers to an enzyme that breaks peptide bonds between amino acids of proteins.
Tertiary structure	Tertiary structure refers to the complex three-dimensional structure of a single peptide chain; held in place by disulfide bonds between cysteines.
Denaturation	Denaturation is a structural change in biomolecules such as nucleic acids and proteins, such that they are no longer in their native state, and their shape which allows for optimal activity.
Fruit	A fruit is the ripened ovary—together with seeds—of a flowering plant. In many species, the fruit incorporates the ripened ovary and surrounding tissues.
Starch	Biochemically, starch is a combination of two polymeric carbohydrates (polysaccharides) called amylose and amylopectin.
Photosynthesis	Photosynthesis is a biochemical process in which plants, algae, and some bacteria harness the energy of light to produce food. Ultimately, nearly all living things depend on energy produced from photosynthesis for their nourishment, making it vital to life on Earth.
Monosaccharide	A monosaccharide is simplest form of a carbohydrate. They consist of one sugar and are usually colorless, water-soluble, crystalline solids. Some monosaccharides have a sweet taste. They are the building blocks of disaccharides like sucrose and polysaccharides.
Glucose	Glucose, a simple monosaccharide sugar, is one of the most important carbohydrates and is used as a source of energy in animals and plants. Glucose is one of the main products of

photosynthesis and starts respiration.

Isomer	An isomer is a molecule with the same chemical formula and often with the same kinds of bonds between atoms, but in which the atoms are arranged differently. That is to say, they have different structural formula.
Ribose	Ribose is an aldopentose — a monosaccharide containing five carbon atoms, and including an aldehyde functional group. It has chemical formula $C_5H_{10}O_5$.
Disaccharide	A disaccharide is a sugar (a carbohydrate) composed of two monosaccharides. The two monosaccharides are bonded via a condensation reaction.
Linkage	Linkage refers to the patterns of assortment of genes that are located on the same chromosome. Important because if the genes are located relatively far apart, crossing over is more likely to occur between them than if they are close together.
Microorganism	A microorganism is an organism that is so small that it is microscopic (invisible to the naked eye). They are often illustrated using single-celled, or unicellular organisms; however, some unicellular protists are visible to the naked eye, and some multicellular species are microscopic.
Blood	Blood is a circulating tissue composed of fluid plasma and cells. The main function of blood is to supply nutrients (oxygen, glucose) and constitutional elements to tissues and to remove waste products.
Polysaccharide	Polymer made from sugar monomers is a polysaccharide. They are relatively complex carbohydrates.
Enzyme	An enzyme is a protein that catalyzes, or speeds up, a chemical reaction. They are essential to sustain life because most chemical reactions in biological cells would occur too slowly, or would lead to different products, without them.
Glycogen	Glycogen refers to a complex, extensively branched polysaccharide of many glucose monomers; serves as an energy-storage molecule in liver and muscle cells.
Cartilage	Cartilage is a type of dense connective tissue. Cartilage is composed of cells called chondrocytes which are dispersed in a firm gel-like ground substance, called the matrix. Cartilage is avascular (contains no blood vessels) and nutrients are diffused through the matrix.
Polymer	Polymer is a generic term used to describe a very long molecule consisting of structural units and repeating units connected by covalent chemical bonds.
Species	Group of similarly constructed organisms capable of interbreeding and producing fertile offspring is a species.
Pentose	A pentose is a monosaccharide with five carbon atoms.
Guanine	Guanine is one of the five main nucleobases found in nucleic acids. Guanine is a purine derivative, and in Watson-Crick base pairing forms three hydrogen bonds with cytosine. Guanine "stacks" vertically with the other nucleobases via aromatic interactions.
Cytosine	Cytosine is one of the 5 main nucleobases used in storing and transporting genetic information within a cell in the nucleic acids DNA and RNA. It is a pyrimidine derivative, with a heterocyclic aromatic ring and two substituents attached. The nucleoside of cytosine is cytidine.
Purine	Purine refers to one of two families of nitrogenous bases found in nucleotides. Adenine and guanine are purines.
Vitamin	A Vitamin is an organic molecule required by a living organism in minute amounts for proper

health. An organism deprived of all sources of a particular vitamin will eventually suffer from disease symptoms specific to that vitamin.

Glycerol

Glycerol is a three-carbon substance that forms the backbone of fatty acids in fats. When the body uses stored fat as a source of energy, glycerol and fatty acids are released into the bloodstream. The glycerol component can be converted to glucose by the liver and provides energy for cellular metabolism.

Hydrocarbon

A chemical compound composed only of the elements carbon and hydrogen is called hydrocarbon.

Triglyceride

Triglyceride is a glyceride in which the glycerol is esterified with three fatty acids. They are the main constituent of vegetable oil and animal fats and play an important role in metabolism as energy sources. They contain a bit more than twice as much energy as carbohydrates and proteins.

Double bond

Double bond refers to a type of covalent bond in which two atoms share two pairs of electrons; symbolized by a pair of lines between the bonded atoms. An example is in ethylene (between the carbon atoms). It usually consists of one sigma bond and one pi bond.

Ion

Ion refers to an atom or molecule that has gained or lost one or more electrons, thus acquiring an electrical charge.

Egg

An egg is the zygote, resulting from fertilization of the ovum. It nourishes and protects the embryo.

Steroid

A steroid is a lipid characterized by a carbon skeleton with four fused rings. Different steroids vary in the functional groups attached to these rings. Hundreds of distinct steroids have been identified in plants and animals. Their most important role in most living systems is as hormones.

Estrogen

Estrogen is a steroid that functions as the primary female sex hormone. While present in both men and women, they are found in women in significantly higher quantities.

Digestion

Digestion refers to the mechanical and chemical breakdown of food into molecules small enough for the body to absorb; the second main stage of food processing, following ingestion.

Liver

The liver is an organ in vertebrates, including humans. It plays a major role in metabolism and has a number of functions in the body including drug detoxification, glycogen storage, and plasma protein synthesis. It also produces bile, which is important for digestion.

Absorption

Absorption is a physical or chemical phenomenon or a process in which atoms, molecules, or ions enter some bulk phase - gas, liquid or solid material. In nutrition, amino acids are broken down through digestion, which begins in the stomach.

Oxidation

Oxidation refers to the loss of electrons from a substance involved in a redox reaction; always accompanies reduction.

Bacteria

The domain that contains procaryotic cells with primarily diacyl glycerol diesters in their membranes and with bacterial rRNA. Bacteria also is a general term for organisms that are composed of procaryotic cells and are not multicellular.

Skin

Skin is an organ of the integumentary system composed of a layer of tissues that protect underlying muscles and organs.

Carrier protein

Protein molecule that combines with a substance and transports it through the plasma membrane is called a carrier protein.

Bacteria	The domain that contains procaryotic cells with primarily diacyl glycerol diesters in their membranes and with bacterial rRNA. Bacteria also is a general term for organisms that are composed of procaryotic cells and are not multicellular.
Antibiotic	Antibiotic refers to substance such as penicillin or streptomycin that is toxic to microorganisms. Usually a product of a particular microorvanism or plant.
Cell	The cell is the structural and functional unit of all living organisms, and is sometimes called the "building block of life."
Algae	The algae consist of several different groups of living organisms that capture light energy through photosynthesis, converting inorganic substances into simple sugars with the captured energy.
Multicellular	Multicellular organisms are those organisms consisting of more than one cell, and having differentiated cells that perform specialized functions. Most life that can be seen with the naked eye is multicellular, as are all animals (i.e. members of the kingdom Animalia) and plants (i.e. members of the kingdom Plantae).
Light microscope	An optical instrument with lenses that refract visible light to magnify images and project them into a viewer's eye or onto photographic film is referred to as light microscope.
Microscope	A microscope is an instrument for viewing objects that are too small to be seen by the naked or unaided eye.
Electron microscope	The electron microscope is a microscope that can magnify very small details with high resolving power due to the use of electrons as the source of illumination, magnifying at levels up to 500,000 times.
Electron	The electron is a light fundamental subatomic particle that carries a negative electric charge. The electron is a spin-1/2 lepton, does not participate in strong interactions and has no substructure.
Plasma membrane	Membrane surrounding the cytoplasm that consists of a phospholipid bilayer with embedded proteins is referred to as plasma membrane.
Plasma	In physics and chemistry, a plasma is an ionized gas, and is usually considered to be a distinct phase of matter. "Ionized" in this case means that at least one electron has been dissociated from a significant fraction of the molecules.
Cytoplasm	Cytoplasm refers to everything inside a cell between the plasma membrane and the nucleus; consists of a semifluid medium and organelles.
Eukaryotic cell	Eukaryotic cell refers to a type of cell that has a membrane-enclosed nucleus and other membrane enclosed organelles. All organisms except bacteria and archaea are composed of eukaryotic cells.
Evolution	In biology, evolution is the process by which novel traits arise in populations and are passed on from generation to generation. Its action over large stretches of time explains the origin of new species and ultimately the vast diversity of the biological world.
Prokaryote	A prokaryote is an organisms without a cell nucleus, or indeed any other membrane-bound organelles, in most cases unicellular (in rare cases, multicellular). This is in contrast to a eukaryote, organisms that have cell nuclei and may be variously unicellular or multicellular.
Nucleoid	Central region of a bacterium where the bacterial chromosome is found is referred to as a nucleoid.
Granule	A granule is a small grain. In describing cells, a granule may be any structure barely visible by light microscopy, but most often a secretory vesicle.

Go to **Cram101.com** for the Practice Tests for this Chapter.

Prokaryotic cell	A cell lacking a membrane-bounded nucleus and organelles is referred to as a prokaryotic cell.
Cell wall	Cell wall refers to a protective layer external to the plasma membrane in plant cells, bacteria, fungi, and some protists; protects the cell and helps maintain its shape.
Capsule	A sticky layer that surrounds the bacterial cell wall, protects the cell surface, and sometimes helps glue the cell to surfaces is called the capsule. In botany, a capsule is a type of dry fruit as in the poppy, iris, foxglove, etc. as well as another term for the sporangium of mosses and hornworts.
Flagella	Flagella are whip-like organelle that many unicellular organisms, and some multicellular ones, use to move about.
Cell division	Cell division is the process by which a cell (called the parent cell) divides into two cells (called daughter cells). Cell division is usually a small segment of a larger cell cycle. In meiosis, however, a cell is permanently transformed and cannot divide again.
Flagellum	A flagellum is a whip-like organelle that many unicellular organisms, and some multicellular ones, use to move about.
Nucleus	In cell biology, the nucleus is found in all eukaryotic cells that contains most of the cell's genetic material. The nucleus has two primary functions: to control chemical reactions within the cytoplasm and to store information needed for cellular division.
Mitochondrion	Mitochondrion refers to an organelle in eukaryotic cells where cellular respiration occurs. Enclosed by two concentric membranes, it is where most of the cell's ATP is made.
Endoplasmic reticulum	The endoplasmic reticulum is an organelle found in all eukaryotic cells. It modifies proteins, makes macromolecules, and transfers substances throughout the cell.
Lysosome	Lysosome refers to a digestive organelle in eukaryotic cells; contains hydrolytic enzymes that digest the cell's food and wastes. They are found in both plant and animal cells, and are built in the Golgi apparatus.
Nuclear envelope	The nuclear envelope refers to the double membrane of the nucleus that encloses genetic material in eukaryotic cells. It separates the contents of the nucleus (DNA in particular) from the cytosol.
Protein	A protein is a complex, high-molecular-weight organic compound that consists of amino acids joined by peptide bonds. They are essential to the structure and function of all living cells and viruses. Many are enzymes or subunits of enzymes.
Chromatin	Chromatin refers to the combination of DNA and proteins that constitute chromosomes; often used to refer to the diffuse, very extended form taken by the chromosomes when a eukaryotic cell is not dividing.
Nucleoplasm	Semifluid medium of the nucleus, containing chromatin is called nucleoplasm.
Ribosome	A ribosome is an organelle composed of rRNA and ribosomal proteins. It translates mRNA into a polypeptide chain (e.g., a protein). It can be thought of as a factory that builds a protein from a set of genetic instructions.
Endomembrane system	The endomembrane system is the system of internal membranes within eukaryotic cells that divide the cell into functional and structural compartments, or organelles.
Vesicle	In cell biology, a vesicle is a relatively small and enclosed compartment, separated from the cytosol by at least one lipid bilayer.
Carbohydrate	Carbohydrate is a chemical compound that contains oxygen, hydrogen, and carbon atoms. They consist of monosaccharide sugars of varying chain lengths and that have the general chemical

formula $C_n(H_2O)_n$ or are derivatives of such.

Nerve	A nerve is an enclosed, cable-like bundle of nerve fibers or axons, which includes the glia that ensheath the axons in myelin.
Golgi apparatus	Golgi apparatus refers to an organelle in eukaryotic cells consisting of stacks of membranous sacs that modify, store, and ship products of the endoplasmic reticulum.
Golgi	Golgi discovered a method of staining nervous tissue which would stain a limited number of cells at random, in their entirety. This enabled him to view the paths of nerve cells in the brain for the first time. He called his discovery the black reaction. It is now known universally as the Golgi stain.
Invertebrate	Invertebrate is a term coined by Jean-Baptiste Lamarck to describe any animal without a spinal column. It therefore includes all animals except vertebrates (fish, reptiles, amphibians, birds and mammals).
Vertebrate	Vertebrate is a subphylum of chordates, specifically, those with backbones or spinal columns. They started to evolve about 530 million years ago during the Cambrian explosion, which is part of the Cambrian period.
Organelle	In cell biology, an organelle is one of several structures with specialized functions, suspended in the cytoplasm of a eukaryotic cell.
Phagocytosis	Phagocytosis is a form of endocytosis where large particles are enveloped by the cell membrane of a (usually larger) cell and internalized to form a phagosome, or "food vacuole."
Enzyme	An enzyme is a protein that catalyzes, or speeds up, a chemical reaction. They are essential to sustain life because most chemical reactions in biological cells would occur too slowly, or would lead to different products, without them.
Central vacuole	A membrane-enclosed sac occupying most of the interior of a mature plant cell, having diverse roles in reproduction, growth, and development is called central vacuole.
Vacuole	A vacuole is a large membrane-bound compartment within some eukaryotic cells where they serve a variety of different functions: capturing food materials or unwanted structural debris surrounding the cell, sequestering materials that might be toxic to the cell, maintaining fluid balance (called turgor) within the cell.
Matrix	In biology, matrix (plural: matrices) is the material between animal or plant cells, the material (or tissue) in which more specialized structures are embedded, and a specific part of the mitochondrion that is the site of oxidation of organic molecules.
Liver	The liver is an organ in vertebrates, including humans. It plays a major role in metabolism and has a number of functions in the body including drug detoxification, glycogen storage, and plasma protein synthesis. It also produces bile, which is important for digestion.
Pigment	Pigment is any material resulting in color in plant or animal cells which is the result of selective absorption.
Photosynthesis	Photosynthesis is a biochemical process in which plants, algae, and some bacteria harness the energy of light to produce food. Ultimately, nearly all living things depend on energy produced from photosynthesis for their nourishment, making it vital to life on Earth.
Mitochondrial matrix	Mitochondrial matrix refers to the fluid contained within the inner membrane of a mitochondrion.
Leaf	In botany, a leaf is an above-ground plant organ specialized for photosynthesis. For this purpose, a leaf is typically flat (laminar) and thin, to expose the chloroplast containing cells (chlorenchyma tissue) to light over a broad area, and to allow light to penetrate fully into the tissues.

Go to **Cram101.com** for the Practice Tests for this Chapter.
And, **NEVER** highlight a book again!

Flower	A flower is the reproductive structure of a flowering plant. The flower structure contains the plant's reproductive organs, and its function is to produce seeds through sexual reproduction.
Starch	Biochemically, starch is a combination of two polymeric carbohydrates (polysaccharides) called amylose and amylopectin.
Lipid	Lipid is one class of aliphatic hydrocarbon-containing organic compounds essential for the structure and function of living cells. They are characterized by being water-insoluble but soluble in nonpolar organic solvents.
Petal	A petal is one member or part of the corolla of a flower. It is the inner part of the perianth that comprises the sterile parts of a flower and consists of inner and outer tepals.
Contractile vacuole	A fluid-filled vacuole in certain protists that takes up excess water from the cytoplasm, contracts, and expels the water outside the cell through a pore in the plasma membrane is a contractile vacuole.
Actin	Actin is a globular protein that polymerizes helically forming filaments, which like the other two components of the cellular cytoskeleton form a three-dimensional network inside an eukaryotic cell. They provide mechanical support for the cell, determine the cell shape, enable cell movements .
Muscle	Muscle is a contractile form of tissue. It is one of the four major tissue types, the other three being epithelium, connective tissue and nervous tissue. Muscle contraction is used to move parts of the body, as well as to move substances within the body.
Intestine	The intestine is the portion of the alimentary canal extending from the stomach to the anus and, in humans and mammals, consists of two segments, the small intestine and the large intestine. The intestine is the part of the body responsible for extracting nutrition from food.
Intermediate filament	Intermediate filament refers to an intermediate-sized protein fiber that is one of the three main kinds of fibers making up the cytoskeleton of eukaryotic cells; ropelike, made of fibrous proteins.
Filament	The stamen is the male organ of a flower. Each stamen generally has a stalk called the filament, and, on top of the filament, an anther. The filament is a long chain of proteins, such as those found in hair, muscle, or in flagella.
Polypeptide	Polypeptide refers to polymer of many amino acids linked by peptide bonds.
Microtubule	Microtubule is a protein structure found within cells, one of the components of the cytoskeleton. They have diameter of ~ 24 nm and varying length from several micrometers to possible millimeters in axons of nerve cells. They serve as structural components within cells and are involved in many cellular processes including mitosis, cytokinesis, and vesicular transport.
Chromosome	A chromosome is, minimally, a very long, continuous piece of DNA, which contains many genes, regulatory elements and other intervening nucleotide sequences.
Cilia	Numerous short, hairlike structures projecting from the cell surface that enable locomotion are cilia.
Basal body	A basal body is an organelle formed from a centriole, a short cylindrical array of microtubules. It is found at the base of a eukaryotic cell cilium or flagellum and serves as a nucleation site for the growth of the axoneme microtubules.
Peptidoglycan	Peptidoglycan refers to a polymer of complex sugars cross-linked by short polypeptides; a material unique to eubacterial cell walls.

Go to **Cram101.com** for the Practice Tests for this Chapter.

Cellulose	A large polysaccharide composed of many glucose monomers linked into cable-like fibrils that provide structural support in plant cell walls is referred to as cellulose.
Extracellular matrix	Extracellular matrix is any material part of a tissue that is not part of any cell. Extracellular matrix is the defining feature of connective tissue.
Basement membrane	Basement membrane refers to the extracellular matrix, consisting of a dense mat of proteins and sticky polysaccharides, that anchors an epithelium to underlying tissues.

Go to **Cram101.com** for the Practice Tests for this Chapter.

Go to **Cram101.com** for the Practice Tests for this Chapter.
And, **NEVER** highlight a book again!

Skin	Skin is an organ of the integumentary system composed of a layer of tissues that protect underlying muscles and organs.
Cell	The cell is the structural and functional unit of all living organisms, and is sometimes called the "building block of life."
Phospholipid	Phospholipid is a class of lipids formed from four components: fatty acids, a negatively-charged phosphate group, an alcohol and a backbone. Phospholipids with a glycerol backbone are known as glycerophospholipids or phosphoglycerides.
Lipid	Lipid is one class of aliphatic hydrocarbon-containing organic compounds essential for the structure and function of living cells. They are characterized by being water-insoluble but soluble in nonpolar organic solvents.
Fatty acid	A fatty acid is a carboxylic acid (or organic acid), often with a long aliphatic tail (long chains), either saturated or unsaturated.
Acid	An acid is a water-soluble, sour-tasting chemical compound that when dissolved in water, gives a solution with a pH of less than 7.
Hydrophobic	Hydrophobic refers to being electrically neutral and nonpolar, and thus prefering other neutral and nonpolar solvents or molecular environments. Hydrophobic is often used interchangeably with "oily" or "lipophilic."
Molecule	A molecule is the smallest particle of a pure chemical substance that still retains its chemical composition and properties.
Organelle	In cell biology, an organelle is one of several structures with specialized functions, suspended in the cytoplasm of a eukaryotic cell.
Plasma	In physics and chemistry, a plasma is an ionized gas, and is usually considered to be a distinct phase of matter. "Ionized" in this case means that at least one electron has been dissociated from a significant fraction of the molecules.
Hydrophilic	A hydrophilic molecule or portion of a molecule is one that is typically charge-polarized and capable of hydrogen bonding, enabling it to dissolve more readily in water than in oil or other hydrophobic solvents.
Electron	The electron is a light fundamental subatomic particle that carries a negative electric charge. The electron is a spin-1/2 lepton, does not participate in strong interactions and has no substructure.
Fluid mosaic	The basic composition and structure of the plasma membrane is the same as that of the membranes that surround organelles and other subcellular compartments. The foundation is a phospholipid bilayer, and the membrane as a whole is often described as a fluid mosaic – a two-dimensional fluid of freely diffusing lipids, dotted or embedded with proteins, which may function as channels or transporters across the membrane, or as receptors.
Mosaic	An organism containing tissues of different genotypes is referred to as mosaic.
Carbohydrate	Carbohydrate is a chemical compound that contains oxygen, hydrogen, and carbon atoms. They consist of monosaccharide sugars of varying chain lengths and that have the general chemical formula $C_n(H_2O)_n$ or are derivatives of such.
Monosaccharide	A monosaccharide is simplest form of a carbohydrate. They consist of one sugar and are usually colorless, water-soluble, crystalline solids. Some monosaccharides have a sweet taste. They are the building blocks of disaccharides like sucrose and polysaccharides.
Sponge	An invertebrates that consist of a complex aggregation of cells, including collar cells, and has a skeleton of fibers and/or spicules is a sponge. They are primitive, sessile, mostly marine, waterdwelling filter feeders that pump water through their matrix to filter out

Go to **Cram101.com** for the Practice Tests for this Chapter.

particulates of food matter.

Species	Group of similarly constructed organisms capable of interbreeding and producing fertile offspring is a species.
Tissue	Group of similar cells which perform a common function is called tissue.
Sperm	Sperm refers to the male sex cell with three distinct parts at maturity: head, middle piece, and tail.
Flagella	Flagella are whip-like organelle that many unicellular organisms, and some multicellular ones, use to move about.
Plasma membrane	Membrane surrounding the cytoplasm that consists of a phospholipid bilayer with embedded proteins is referred to as plasma membrane.
Adhesion	The molecular attraction exerted between the surfaces of unlike bodies in contact, as water molecules to the walls of the narrow tubes that occur in plants is referred to as adhesion.
Multicellular	Multicellular organisms are those organisms consisting of more than one cell, and having differentiated cells that perform specialized functions. Most life that can be seen with the naked eye is multicellular, as are all animals (i.e. members of the kingdom Animalia) and plants (i.e. members of the kingdom Plantae).
Desmosome	A strong cell-tocell junction that attaches adjacent cells to one another is referred to as desmosome.
Gap junction	A gap junction is a junction between certain animal/plant cell-types that allows different molecules and ions to pass freely between cells. The junction connects the cytoplasm of cells.
Diffusion	Diffusion refers to the spontaneous movement of particles of any kind from where they are more concentrated to where they are less concentrated.
Pigment	Pigment is any material resulting in color in plant or animal cells which is the result of selective absorption.
Net movement	Net movement refers to movement in one direction minus the movement in the other.
Ion	Ion refers to an atom or molecule that has gained or lost one or more electrons, thus acquiring an electrical charge.
Concentration gradient	Gradual change in chemical concentration from one point to another is called concentration gradient.
Gradient	Gradient refers to a difference in concentration, pressure, or electrical charge between two regions.
Solute	Substance that is dissolved in a solvent, forming a solution is referred to as a solute.
Blood	Blood is a circulating tissue composed of fluid plasma and cells. The main function of blood is to supply nutrients (oxygen, glucose) and constitutional elements to tissues and to remove waste products.
Hypertonic solution	A solution that has a higher concentration of solutes than that in a cell is said to be a hypertonic solution. This solution has more solute particles and, therefore, relatively less water than the cell contents.
Hypertonic	A hypertonic cell environment has a higher concentration of solutes than in cytoplasm. In a hypertonic environment, osmosis causes water to flow out of the cell. If enough water is removed in this way, the cytoplasm will have such a small concentration of water that the cell has difficulty functioning.

31

Hypotonic solution	A hypotonic solution contains a higher concentration compared to the cell.
Hypotonic	In comparing two solutions, referring to the one with the lower concentration of solutes is hypotonic.
Red blood cell	The red blood cell is the most common type of blood cell and is the vertebrate body's principal means of delivering oxygen from the lungs or gills to body tissues via the blood.
Archaea	The Archaea are a major division of living organisms. Although there is still uncertainty in the exact phylogeny of the groups, Archaea, Eukaryotes and Bacteria are the fundamental classifications in what is called the three-domain system.
Protein	A protein is a complex, high-molecular-weight organic compound that consists of amino acids joined by peptide bonds. They are essential to the structure and function of all living cells and viruses. Many are enzymes or subunits of enzymes.
Facilitated diffusion	Facilitated diffusion is a process of diffusion, a form of passive transport, via which molecules diffuse across membranes, with the assistance of transport proteins.
Glucose	Glucose, a simple monosaccharide sugar, is one of the most important carbohydrates and is used as a source of energy in animals and plants. Glucose is one of the main products of photosynthesis and starts respiration.
Carrier protein	Protein molecule that combines with a substance and transports it through the plasma membrane is called a carrier protein.
Active transport	Active transport is the mediated transport of biochemicals, and other atomic/molecular substances, across membranes. In this form of transport, molecules move against either an electrical or concentration gradient.
Intestine	The intestine is the portion of the alimentary canal extending from the stomach to the anus and, in humans and mammals, consists of two segments, the small intestine and the large intestine. The intestine is the part of the body responsible for extracting nutrition from food.
Nucleic acid	A nucleic acid is a complex, high-molecular-weight biochemical macromolecule composed of nucleotide chains that convey genetic information. The most common are deoxyribonucleic acid (DNA) and ribonucleic acid (RNA). They are found in all living cells and viruses.
Phagocytosis	Phagocytosis is a form of endocytosis where large particles are enveloped by the cell membrane of a (usually larger) cell and internalized to form a phagosome, or "food vacuole."
Unicellular	Unicellular organisms carry out all the functions of life. Unicellular species are those whose members consist of a single cell throughout their life cycle. This latter qualification is significant since most multicellular organisms consist of a single cell at the beginning of their life cycles.
Food vacuole	The simplest type of digestive cavity, found in single-celled organisms is a food vacuole.
Vacuole	A vacuole is a large membrane-bound compartment within some eukaryotic cells where they serve a variety of different functions: capturing food materials or unwanted structural debris surrounding the cell, sequestering materials that might be toxic to the cell, maintaining fluid balance (called turgor) within the cell.
Receptor-mediated endocytosis	Receptor-mediated endocytosis refers to the movement of specific molecules into a cell by the inward budding of membranous vesicles. The vesicles contain proteins with receptor sites specific to the molecules being taken in.
Endocytosis	Endocytosis is a process where cells absorb material (molecules or other cells) from outside by engulfing it with their cell membranes.

Go to **Cram101.com** for the Practice Tests for this Chapter.

Receptor protein	Protein located in the plasma membrane or within the cell that binds to a substance that alters some metabolic aspect of the cell is referred to as receptor protein. It will only link up with a substance that has a certain shape that allows it to bind to the receptor.
Receptor	A receptor is a protein on the cell membrane or within the cytoplasm or cell nucleus that binds to a specific molecule (a ligand), such as a neurotransmitter, hormone, or other substance, and initiates the cellular response to the ligand. Receptor, in immunology, the region of an antibody which shows recognition of an antigen.
Vesicle	In cell biology, a vesicle is a relatively small and enclosed compartment, separated from the cytosol by at least one lipid bilayer.
Liver	The liver is an organ in vertebrates, including humans. It plays a major role in metabolism and has a number of functions in the body including drug detoxification, glycogen storage, and plasma protein synthesis. It also produces bile, which is important for digestion.
Secretion	Secretion is the process of segregating, elaborating, and releasing chemicals from a cell, or a secreted chemical substance or amount of substance.
Ribosome	A ribosome is an organelle composed of rRNA and ribosomal proteins. It translates mRNA into a polypeptide chain (e.g., a protein). It can be thought of as a factory that builds a protein from a set of genetic instructions.
Nerve	A nerve is an enclosed, cable-like bundle of nerve fibers or axons, which includes the glia that ensheath the axons in myelin.
Enzyme	An enzyme is a protein that catalyzes, or speeds up, a chemical reaction. They are essential to sustain life because most chemical reactions in biological cells would occur too slowly, or would lead to different products, without them.
Smooth endoplasmic reticulum	Smooth endoplasmic reticulum refers to endoplasmic reticulum without ribosomes. It has functions in several metabolic processes, including synthesis of lipids, metabolism of carbohydrates, and detoxification of drugs and poisons.
Endoplasmic reticulum	The endoplasmic reticulum is an organelle found in all eukaryotic cells. It modifies proteins, makes macromolecules, and transfers substances throughout the cell.
Rough endoplasmic reticulum	Rough endoplasmic reticulum manufactures and transports proteins destined for membranes and secretion. It is called "rough" because ribosomes present on the cytosolic side of the membrane give it a rough appearance.
Eukaryotic cell	Eukaryotic cell refers to a type of cell that has a membrane-enclosed nucleus and other membrane enclosed organelles. All organisms except bacteria and archaea are composed of eukaryotic cells.
Golgi apparatus	Golgi apparatus refers to an organelle in eukaryotic cells consisting of stacks of membranous sacs that modify, store, and ship products of the endoplasmic reticulum.
Golgi	Golgi discovered a method of staining nervous tissue which would stain a limited number of cells at random, in their entirety. This enabled him to view the paths of nerve cells in the brain for the first time. He called his discovery the black reaction. It is now known universally as the Golgi stain.

Leech	The leech is a annelid comprising the subclass Hirudinea. There are freshwater, terrestrial and marine leeches. Like their near relatives, the Oligochaeta, they share the presence of a clitellum.Like earthworms, leeches are hermaphrodites.
Thrombin	Thrombin refers to an enzyme that converts fibrinogen to fibrin threads during blood clotting.
Cell	The cell is the structural and functional unit of all living organisms, and is sometimes called the "building block of life."
Enzyme	An enzyme is a protein that catalyzes, or speeds up, a chemical reaction. They are essential to sustain life because most chemical reactions in biological cells would occur too slowly, or would lead to different products, without them.
Biochemistry	Biochemistry studies how complex chemical reactions give rise to life. It is a hybrid branch of chemistry which specialises in the chemical processes in living organisms.
Kinetic energy	Kinetic energy refers to energy that is actually doing work; the energy of a mass of matter that is moving. Moving matter performs work by transferring its motion to other matter, such as leg muscles pushing bicycle pedals.
Potential energy	Stored energy as a result of location or spatial arrangement is referred to as potential energy.
Chemical energy	Chemical energy refers to energy stored in the chemical bonds of molecules; a form of potential energy.
Protein	A protein is a complex, high-molecular-weight organic compound that consists of amino acids joined by peptide bonds. They are essential to the structure and function of all living cells and viruses. Many are enzymes or subunits of enzymes.
Active transport	Active transport is the mediated transport of biochemicals, and other atomic/molecular substances, across membranes. In this form of transport, molecules move against either an electrical or concentration gradient.
Free energy	The term thermodynamic free energy denotes the total amount of energy in a physical system which can be converted to do work.
Chemical reaction	Chemical reaction refers to a process leading to chemical changes in matter; involves the making and/or breaking of chemical bonds.
Entropy	Entropy is a measure of the amount of energy in a physical system that cannot be used to do work. In simpler terms, it is also a measure of the disorder and randomness present in a system.
Glucose	Glucose, a simple monosaccharide sugar, is one of the most important carbohydrates and is used as a source of energy in animals and plants. Glucose is one of the main products of photosynthesis and starts respiration.
Reactant	A reactant is any substance initially present in a chemical reaction. These reactants react with each other to form the products of a chemical reaction. In a chemical equation, the reactants are the elements or compounds on the left hand side of the reaction equation.
Exergonic	Exergonic means to release energy in the form of work. By thermodynamic standards, work, a form of energy, is defined normally to move from the system (the internal region) to the surroundings (the external region).
Hydrolysis	Hydrolysis is a chemical process in which a molecule is cleaved into two parts by the addition of a molecule of water.
Amino acid	An amino acid is any molecule that contains both amino and carboxylic acid functional groups.

Go to **Cram101.com** for the Practice Tests for this Chapter.

	They are the basic structural building units of proteins. They form short polymer chains called peptides or polypeptides which in turn form structures called proteins.
Acid	An acid is a water-soluble, sour-tasting chemical compound that when dissolved in water, gives a solution with a pH of less than 7.
Coupled reaction	Coupled reaction refers to a pair of reactions, one cxergonic and one endergonic, that are linked together such that the energy produced by the exergonic reaction provides the energy needed to drive the endergonic reaction.
Evolution	In biology, evolution is the process by which novel traits arise in populations and are passed on from generation to generation. Its action over large stretches of time explains the origin of new species and ultimately the vast diversity of the biological world.
Species	Group of similarly constructed organisms capable of interbreeding and producing fertile offspring is a species.
Activation energy	The activation energy is the threshold energy, or the energy that must be overcome in order for a chemical reaction to occur. Activation energy may otherwise be denoted as the minimum energy necessary for a specific chemical reaction to occur.
Functional group	Functional group is a submolecular structural motif, characterized by specific elemental composition and connectivity, that confer reactivity upon the molecule that contains it.
Enzyme-substrate complex	A temporary molecule formed when an enzyme attaches itself to a substrate molecule is an enzyme-substrate complex.
Substrate	A substrate is a molecule which is acted upon by an enzyme. Each enzyme recognizes only the specific substrate of the reaction it catalyzes. A surface in or on which an organism lives.
Blood	Blood is a circulating tissue composed of fluid plasma and cells. The main function of blood is to supply nutrients (oxygen, glucose) and constitutional elements to tissues and to remove waste products.
Peptide bond	A peptide bond is a chemical bond formed between two molecules when the carboxyl group of one molecule reacts with the amino group of the other molecule, releasing a molecule of water.
Peptide	Peptide is the family of molecules formed from the linking, in a defined order, of various amino acids. The link between one amino acid residue and the next is an amide bond, and is sometimes referred to as a peptide bond.
Active site	The active site of an enzyme is the binding site where catalysis occurs. The structure and chemical properties of the active site allow the recognition and binding of the substrate.
Catalysis	Catalysis is the acceleration of the reaction rate of a chemical reaction by means of a substance, called a catalyst, that is itself not consumed by the overall reaction.
Ion	Ion refers to an atom or molecule that has gained or lost one or more electrons, thus acquiring an electrical charge.
Carbohydrate	Carbohydrate is a chemical compound that contains oxygen, hydrogen, and carbon atoms. They consist of monosaccharide sugars of varying chain lengths and that have the general chemical formula $C_n(H_2O)_n$ or are derivatives of such.
Coenzyme	Nonprotein organic molecule that aids the action of the enzyme to which it is loosely bound is referred to as coenzyme.
Vitamin	A Vitamin is an organic molecule required by a living organism in minute amounts for proper health. An organism deprived of all sources of a particular vitamin will eventually suffer from disease symptoms specific to that vitamin.

Go to **Cram101.com** for the Practice Tests for this Chapter.

Homeostasis	Homeostasis is the property of an open system, especially living organisms, to regulate its internal environment to maintain a stable, constant condition, by means of multiple dynamic equilibrium adjustments, controlled by interrelated regulation mechanisms.
Metabolism	Metabolism is the biochemical modification of chemical compounds in living organisms and cells. This includes the biosynthesis of complex organic molecules (anabolism) and their breakdown (catabolism).
Inhibitor	An inhibitor is a type of effector (biology) that decreases or prevents the rate of a chemical reaction. They are often called negative catalysts.
Nerve	A nerve is an enclosed, cable-like bundle of nerve fibers or axons, which includes the glia that ensheath the axons in myelin.
Molecule	A molecule is the smallest particle of a pure chemical substance that still retains its chemical composition and properties.
Hormone	A hormone is a chemical messenger from one cell to another. All multicellular organisms produce hormones. The best known hormones are those produced by endocrine glands of vertebrate animals, but hormones are produced by nearly every organ system and tissue type in a human or animal body. Hormone molecules are secreted directly into the bloodstream, they move by circulation or diffusion to their target cells, which may be nearby cells in the same tissue or cells of a distant organ of the body.
Amylase	Amylase is a digestive enzyme classified as a saccharidase. It is mainly a constituent of pancreatic juice and saliva, needed for the breakdown of long-chain carbohydrates (such as starch) into smaller units.
Acetylcholin-sterase	Acetylcholinesterase is found primarily in the blood and neural synapses. It catalyzes the hydrolysis of the neurotransmitter acetylcholine into choline and acetic acid, a reaction necessary to allow a cholinergic neuron to return to its resting state after activation.

Sugar	A sugar is the simplest molecule that can be identified as a carbohydrate. These include monosaccharides and disaccharides, trisaccharides and the oligosaccharides. The term "glyco-" indicates the presence of a sugar in an otherwise non-carbohydrate substance.
Cell	The cell is the structural and functional unit of all living organisms, and is sometimes called the "building block of life."
Organelle	In cell biology, an organelle is one of several structures with specialized functions, suspended in the cytoplasm of a eukaryotic cell.
Enzyme	An enzyme is a protein that catalyzes, or speeds up, a chemical reaction. They are essential to sustain life because most chemical reactions in biological cells would occur too slowly, or would lead to different products, without them.
Glucose	Glucose, a simple monosaccharide sugar, is one of the most important carbohydrates and is used as a source of energy in animals and plants. Glucose is one of the main products of photosynthesis and starts respiration.
Electron	The electron is a light fundamental subatomic particle that carries a negative electric charge. The electron is a spin-1/2 lepton, does not participate in strong interactions and has no substructure.
Molecule	A molecule is the smallest particle of a pure chemical substance that still retains its chemical composition and properties.
Metabolism	Metabolism is the biochemical modification of chemical compounds in living organisms and cells. This includes the biosynthesis of complex organic molecules (anabolism) and their breakdown (catabolism).
Ion	Ion refers to an atom or molecule that has gained or lost one or more electrons, thus acquiring an electrical charge.
Cellular respiration	Cellular respiration is the process in which the chemical bonds of energy-rich molecules such as glucose are converted into energy usable for life processes.
Respiration	Respiration is the process by which an organism obtains energy by reacting oxygen with glucose to give water, carbon dioxide and ATP (energy). Respiration takes place on a cellular level in the mitochondria of the cells and provide the cells with energy.
Pyruvate	Pyruvate is the ionized form of pyruvic acid. It is an important chemical compound in biochemistry. It is the output of the breakdown of glucose known as glycolysis, and (in aerobic respiration) the main input for the citric acid cycle via acetyl-CoA.
Glycolysis	Glycolysis refers to the multistep chemical breakdown of a molecule of glucose into two molecules of pyruvic acid; the first stage of cellular respiration in all organisms; occurs in the cytoplasmic fluid.
Oxidation	Oxidation refers to the loss of electrons from a substance involved in a redox reaction; always accompanies reduction.
Citric acid cycle	In aerobic organisms, the citric acid cycle is a metabolic pathway that forms part of the break down of carbohydrates, fats and proteins into carbon dioxide and water in order to generate energy. It is the second of three metabolic pathways that are involved in fuel molecule catabolism and ATP production, the other two being glycolysis and oxidative phosphorylation.
Acid	An acid is a water-soluble, sour-tasting chemical compound that when dissolved in water, gives a solution with a pH of less than 7.
Free energy	The term thermodynamic free energy denotes the total amount of energy in a physical system which can be converted to do work.

Exergonic	Exergonic means to release energy in the form of work. By thermodynamic standards, work, a form of energy, is defined normally to move from the system (the internal region) to the surroundings (the external region).
Anaerobic	An anaerobic organism is any organism that does not require oxygen for growth.
Mitochondrion	Mitochondrion refers to an organelle in eukaryotic cells where cellular respiration occurs. Enclosed by two concentric membranes, it is where most of the cell's ATP is made.
Acetyl	The acetyl radical contains a methyl group single-bonded to a carbonyl. The carbon of the carbonyl has an lone electron available, with which it forms a chemical bond to the remainder of the molecule.
Proton	Positive subatomic particle, located in the nucleus and having a weight of approximately one atomic mass unit is referred to as a proton.
Protein	A protein is a complex, high-molecular-weight organic compound that consists of amino acids joined by peptide bonds. They are essential to the structure and function of all living cells and viruses. Many are enzymes or subunits of enzymes.
Active transport	Active transport is the mediated transport of biochemicals, and other atomic/molecular substances, across membranes. In this form of transport, molecules move against either an electrical or concentration gradient.
Potential energy	Stored energy as a result of location or spatial arrangement is referred to as potential energy.
Mitochondrial matrix	Mitochondrial matrix refers to the fluid contained within the inner membrane of a mitochondrion.
Matrix	In biology, matrix (plural: matrices) is the material between animal or plant cells, the material (or tissue) in which more specialized structures are embedded, and a specific part of the mitochondrion that is the site of oxidation of organic molecules.
Gradient	Gradient refers to a difference in concentration, pressure, or electrical charge between two regions.
Diffusion	Diffusion refers to the spontaneous movement of particles of any kind from where they are more concentrated to where they are less concentrated.
Fermentation	Fermentation is the anaerobic metabolic breakdown of a nutrient molecule, such as glucose, without net oxidation. Fermentation does not release all the available energy in a molecule; it merely allows glycolysis to continue by replenishing reduced coenzymes.
Bacteria	The domain that contains procaryotic cells with primarily diacyl glycerol diesters in their membranes and with bacterial rRNA. Bacteria also is a general term for organisms that are composed of procaryotic cells and are not multicellular.
Lactic acid fermentation	Lactic acid fermentation is a form of anaerobic respiration that occurs in animal cells in the absence of oxygen. Glycolysis occurs normally, producing 2 molecules of ATP, 2 molecules of NADH and 2 molecules of pyruvate but the pyruvate is not metabolized to CO_2 in the citric acid cycle.
Lactic acid	Lactic acid accumulates in skeletal muscles during extensive anaerobic exercise, causing temporary muscle pain. Lactic acid is quickly removed from muscles when they resume aerobic metabolism.
Muscle	Muscle is a contractile form of tissue. It is one of the four major tissue types, the other three being epithelium, connective tissue and nervous tissue. Muscle contraction is used to move parts of the body, as well as to move substances within the body.

Lipid	Lipid is one class of aliphatic hydrocarbon-containing organic compounds essential for the structure and function of living cells. They are characterized by being water-insoluble but soluble in nonpolar organic solvents.
Glycerol	Glycerol is a three-carbon substance that forms the backbone of fatty acids in fats. When the body uses stored fat as a source of energy, glycerol and fatty acids are released into the bloodstream. The glycerol component can be converted to glucose by the liver and provides energy for cellular metabolism.
Amino acid	An amino acid is any molecule that contains both amino and carboxylic acid functional groups. They are the basic structural building units of proteins. They form short polymer chains called peptides or polypeptides which in turn form structures called proteins.
Steroid	A steroid is a lipid characterized by a carbon skeleton with four fused rings. Different steroids vary in the functional groups attached to these rings. Hundreds of distinct steroids have been identified in plants and animals. Their most important role in most living systems is as hormones.
Blood	Blood is a circulating tissue composed of fluid plasma and cells. The main function of blood is to supply nutrients (oxygen, glucose) and constitutional elements to tissues and to remove waste products.
Digestive system	The organ system that ingests food, breaks it down into smaller chemical units, and absorbs the nutrient molecules is referred to as the digestive system.
Eukaryotic cell	Eukaryotic cell refers to a type of cell that has a membrane-enclosed nucleus and other membrane enclosed organelles. All organisms except bacteria and archaea are composed of eukaryotic cells.
Cytoskeleton	Cytoskeleton refers to a meshwork of fine fibers in the cytoplasm of a eukaryotic cell; includes microfilaments, intermediate filaments, and microtubules.

Go to **Cram101.com** for the Practice Tests for this Chapter.

Photosynthesis	Photosynthesis is a biochemical process in which plants, algae, and some bacteria harness the energy of light to produce food. Ultimately, nearly all living things depend on energy produced from photosynthesis for their nourishment, making it vital to life on Earth.
Isotope	An isotope is a form of an element whose nuclei have the same atomic number - the number of protons in the nucleus - but different mass numbers because they contain different numbers of neutrons.
Light reactions	The first of two stages in photosynthesis, the steps in which solar energy is absorbed and converted to chemical energy in the form of ATP and NADPH. The light reactions power the sugar-producing Calvin cycle but produce no sugar themselves.
Light reaction	The first of two stages in photosynthesis, the steps in which solar energy is absorbed and converted to chemical energy in the form of ATP and NADPH. The light reaction powers the sugar-producing Calvin cycle but produces no sugar itself.
Chloroplast	A chloroplast is an organelle found in plant cells and eukaryotic algae which conduct photosynthesis. They are similar to mitochondria but are found only in plants. They are surrounded by a double membrane with an intermembrane space; they have their own DNA and are involved in energy metabolism;
Wavelength	The distance between crests of adjacent waves, such as those of the electromagnetic spectrum is wavelength.
Radiation	The emission of electromagnetic waves by all objects warmer than absolute zero is referred to as radiation.
Molecule	A molecule is the smallest particle of a pure chemical substance that still retains its chemical composition and properties.
Pigment	Pigment is any material resulting in color in plant or animal cells which is the result of selective absorption.
Absorption	Absorption is a physical or chemical phenomenon or a process in which atoms, molecules, or ions enter some bulk phase - gas, liquid or solid material. In nutrition, amino acids are broken down through digestion, which begins in the stomach.
Chlorophyll	Chlorophyll is a green photosynthetic pigment found in plants, algae, and cyanobacteria. In plant photosynthesis incoming light is absorbed by chlorophyll and other accessory pigments in the antenna complexes of photosystem I and photosystem II.
Hemoglobin	Hemoglobin is the iron-containing oxygen-transport metalloprotein in the red cells of the blood in mammals and other animals. Hemoglobin transports oxygen from the lungs to the rest of the body, such as to the muscles, where it releases the oxygen load.
Accessory pigment	An accessory pigment is a photosynthetic pigment other than the chlorophylls that enable an organism to utilize more colors of the visible light spectrum for photosynthesis, e.g., carotinoids phycoerythrins and phycocyanin.
Algae	The algae consist of several different groups of living organisms that capture light energy through photosynthesis, converting inorganic substances into simple sugars with the captured energy.
Potential energy	Stored energy as a result of location or spatial arrangement is referred to as potential energy.
Electron	The electron is a light fundamental subatomic particle that carries a negative electric charge. The electron is a spin-1/2 lepton, does not participate in strong interactions and has no substructure.
Reaction center	A photosynthetic reaction center is a protein which is the site of the light reactions of

photosynthesis. The reaction center contains pigments such as chlorophyll and phaeophytin. These absorb light, promoting an electron to a higher energy level within the pigment.

Catabolism	Catabolism is the part of metabolism that partitions molecules into smaller units. It is made up of degradative chemical reactions in the living cell. Large polymeric molecules are processed into their constituent monomeric units.
Free energy	The term thermodynamic free energy denotes the total amount of energy in a physical system which can be converted to do work.
Mitochondrion	Mitochondrion refers to an organelle in eukaryotic cells where cellular respiration occurs. Enclosed by two concentric membranes, it is where most of the cell's ATP is made.
Thylakoid	Flattened sac within a granum whose membrane contains chlorophyll and where the light-dependent reactions of photosynthesis occur is referred to as a thylakoid.
Cell	The cell is the structural and functional unit of all living organisms, and is sometimes called the "building block of life."
Enzyme	An enzyme is a protein that catalyzes, or speeds up, a chemical reaction. They are essential to sustain life because most chemical reactions in biological cells would occur too slowly, or would lead to different products, without them.
Leaf	In botany, a leaf is an above-ground plant organ specialized for photosynthesis. For this purpose, a leaf is typically flat (laminar) and thin, to expose the chloroplast containing cells (chlorenchyma tissue) to light over a broad area, and to allow light to penetrate fully into the tissues.
Glucose	Glucose, a simple monosaccharide sugar, is one of the most important carbohydrates and is used as a source of energy in animals and plants. Glucose is one of the main products of photosynthesis and starts respiration.
Cellular respiration	Cellular respiration is the process in which the chemical bonds of energy-rich molecules such as glucose are converted into energy usable for life processes.
Respiration	Respiration is the process by which an organism obtains energy by reacting oxygen with glucose to give water, carbon dioxide and ATP (energy). Respiration takes place on a cellular level in the mitochondria of the cells and provide the cells with energy.
Bacteria	The domain that contains procaryotic cells with primarily diacyl glycerol diesters in their membranes and with bacterial rRNA. Bacteria also is a general term for organisms that are composed of procaryotic cells and are not multicellular.
Photorespiration	Photorespiration is an alternate pathway for Rubisco, the main enzyme of photosynthesis (specifically, the Calvin cycle). Although Rubisco favors carbon dioxide, it can also use oxygen, producing a glycolate and a glycerate.
Carbohydrate	Carbohydrate is a chemical compound that contains oxygen, hydrogen, and carbon atoms. They consist of monosaccharide sugars of varying chain lengths and that have the general chemical formula $C_n(H_2O)_n$ or are derivatives of such.
Mesophyll	Mesophyll refers to the green tissue in the interior of a leaf; a leaf's ground tissue system, the main site of photosynthesis.
C3 plant	C3 plant refers to a plant that uses the Calvin cycle for the initial steps that incorporate CO_2 into organic material, forming a three-carbon compound as the first stable intermediate.
C4 pathway	C4 pathway refers to the series of reactions in certain plants that fixes carbon dioxide into oxaloacetic acid, which is later broken down for use in the C3 cycle of photosynthesis.

Go to **Cram101.com** for the Practice Tests for this Chapter.

Species	Group of similarly constructed organisms capable of interbreeding and producing fertile offspring is a species.
Metabolism	Metabolism is the biochemical modification of chemical compounds in living organisms and cells. This includes the biosynthesis of complex organic molecules (anabolism) and their breakdown (catabolism).
Acid	An acid is a water-soluble, sour-tasting chemical compound that when dissolved in water, gives a solution with a pH of less than 7.
Glycolysis	Glycolysis refers to the multistep chemical breakdown of a molecule of glucose into two molecules of pyruvic acid; the first stage of cellular respiration in all organisms; occurs in the cytoplasmic fluid.
Pyruvate	Pyruvate is the ionized form of pyruvic acid. It is an important chemical compound in biochemistry. It is the output of the breakdown of glucose known as glycolysis, and (in aerobic respiration) the main input for the citric acid cycle via acetyl-CoA.
Hexose	A hexose is a monosaccharide with six carbon atoms having the chemical formula $C_6H_{12}O_6$.
Lipid	Lipid is one class of aliphatic hydrocarbon-containing organic compounds essential for the structure and function of living cells. They are characterized by being water-insoluble but soluble in nonpolar organic solvents.
Oxidation	Oxidation refers to the loss of electrons from a substance involved in a redox reaction; always accompanies reduction.
Proton	Positive subatomic particle, located in the nucleus and having a weight of approximately one atomic mass unit is referred to as a proton.

Multicellular	Multicellular organisms are those organisms consisting of more than one cell, and having differentiated cells that perform specialized functions. Most life that can be seen with the naked eye is multicellular, as are all animals (i.e. members of the kingdom Animalia) and plants (i.e. members of the kingdom Plantae).
Cell	The cell is the structural and functional unit of all living organisms, and is sometimes called the "building block of life."
Brain	The part of the central nervous system involved in regulating and controlling body activity and interpreting information from the senses transmitted through the nervous system is referred to as the brain.
Tumor	An abnormal mass of cells that forms within otherwise normal tissue is a tumor. This growth can be either malignant or benign
Cell membrane	A component of every biological cell, the selectively permeable cell membrane is a thin and structured bilayer of phospholipid and protein molecules that envelopes the cell. It separates a cell's interior from its surroundings and controls what moves in and out.
Cell division	Cell division is the process by which a cell (called the parent cell) divides into two cells (called daughter cells). Cell division is usually a small segment of a larger cell cycle. In meiosis, however, a cell is permanently transformed and cannot divide again.
Salt	Salt is a term used for ionic compounds composed of positively charged cations and negatively charged anions, so that the product is neutral and without a net charge.
Bacterium	Most bacterium are microscopic and unicellular, with a relatively simple cell structure lacking a cell nucleus, and organelles such as mitochondria and chloroplasts. They are the most abundant of all organisms. They are ubiquitous in soil, water, and as symbionts of other organisms.
Chromosome	A chromosome is, minimally, a very long, continuous piece of DNA, which contains many genes, regulatory elements and other intervening nucleotide sequences.
Plasma membrane	Membrane surrounding the cytoplasm that consists of a phospholipid bilayer with embedded proteins is referred to as plasma membrane.
Plasma	In physics and chemistry, a plasma is an ionized gas, and is usually considered to be a distinct phase of matter. "Ionized" in this case means that at least one electron has been dissociated from a significant fraction of the molecules.
Cytokinesis	The division of the cytoplasm to form two separate daughter cells. Cytokinesis usually occurs during telophase of mitosis, and the two processes make up the mitotic phase of the cell cycle.
Protein	A protein is a complex, high-molecular-weight organic compound that consists of amino acids joined by peptide bonds. They are essential to the structure and function of all living cells and viruses. Many are enzymes or subunits of enzymes.
Reproduction	Biological reproduction is the biological process by which new individual organisms are produced. Reproduction is a fundamental feature of all known life; each individual organism exists as the result of reproduction by an antecedent.
Eukaryotic cell	Eukaryotic cell refers to a type of cell that has a membrane-enclosed nucleus and other membrane enclosed organelles. All organisms except bacteria and archaea are composed of eukaryotic cells.
Nucleus	In cell biology, the nucleus is found in all eukaryotic cells that contains most of the cell's genetic material. The nucleus has two primary functions: to control chemical reactions within the cytoplasm and to store information needed for cellular division.

Gene	Gene refers to a discrete unit of hereditary information consisting of a specific nucleotide sequence in DNA . Most of the genes of a eukaryote are located in its chromosomal DNA; a few are carried by the DNA of mitochondria and chloroplasts.
Duplication	Duplication refers to repetition of part of a chromosome resulting from fusion with a fragment from a homologous chromosome; can result from an error in meiosis or from mutagenesis.
Cytoplasm	Cytoplasm refers to everything inside a cell between the plasma membrane and the nucleus; consists of a semifluid medium and organelles.
Blood	Blood is a circulating tissue composed of fluid plasma and cells. The main function of blood is to supply nutrients (oxygen, glucose) and constitutional elements to tissues and to remove waste products.
Cell cycle	An orderly sequence of events that extends from the time a eukaryotic cell divides to form two daughter cells to the time those daughter cells divide again is called cell cycle.
Interphase	The period in the eukaryotic cell cycle when the cell is not actually dividing is referred to as interphase. During interphase, the cell obtains nutrients, and duplicates its chromosomes.
Root	In vascular plants, the root is that organ of a plant body that typically lies below the surface of the soil. However, this is not always the case, since a root can also be aerial (that is, growing above the ground) or aerating (that is, growing up above the ground or especially above water).
Mitosis	Mitosis is the process by which a cell separates its duplicated genome into two identical halves. It is generally followed immediately by cytokinesis which divides the cytoplasm and cell membrane.
DNA replication	DNA replication is the process of copying a double-stranded DNA strand in a cell, prior to cell division. The two resulting double strands are identical (if the replication went well), and each of them consists of one original and one newly synthesized strand.
Nuclear envelope	The nuclear envelope refers to the double membrane of the nucleus that encloses genetic material in eukaryotic cells. It separates the contents of the nucleus (DNA in particular) from the cytosol.
Cancer	Cancer is a class of diseases or disorders characterized by uncontrolled division of cells and the ability of these cells to invade other tissues, either by direct growth into adjacent tissue through invasion or by implantation into distant sites by metastasis.
White blood cell	The white blood cell is a a component of blood. They help to defend the body against infectious disease and foreign materials as part of the immune system.
Kidney	The kidney is a bean-shaped excretory organ in vertebrates. Part of the urinary system, the kidneys filter wastes (especially urea) from the blood and excrete them, along with water, as urine.
Sperm	Sperm refers to the male sex cell with three distinct parts at maturity: head, middle piece, and tail.
Centromere	The centromere is a region of a eukaryotic chromosome where the kinetochore is assembled; the site where spindle fibers of the mitotic spindle attach to the chromosome during mitosis. It is also the site of the primary constriction visible in microscopy images of chromosomes. Finally, it is the site at which a chromatid and its identical sister attach together during the process of cell reproduction.
Nucleosome	Nucleosome refers to the beadlike unit of DNA packing in a eukaryotic cell; consists of DNA wound around a protein core made up of eight histone molecules.

Chromatin	Chromatin refers to the combination of DNA and proteins that constitute chromosomes; often used to refer to the diffuse, very extended form taken by the chromosomes when a eukaryotic cell is not dividing.
Centrosome	Centrosome refers to material in the cytoplasm of a eukaryotic cell that gives rise to microtubules; important in mitosis and meiosis; also called microtubule-organizing center.
Centriole	A centriole is a barrel shaped microtubule structure found in most animal cells, and cells of fungi and algae though not frequently in plants. The walls of each centriole are usually composed of nine triplet microtubules.
Microtubule	Microtubule is a protein structure found within cells, one of the components of the cytoskeleton. They have diameter of ~ 24 nm and varying length from several micrometers to possible millimeters in axons of nerve cells. They serve as structural components within cells and are involved in many cellular processes including mitosis, cytokinesis, and vesicular transport.
Light microscope	An optical instrument with lenses that refract visible light to magnify images and project them into a viewer's eye or onto photographic film is referred to as light microscope.
Microscope	A microscope is an instrument for viewing objects that are too small to be seen by the naked or unaided eye.
Prophase	Prophase is a stage of mitosis in which chromatin, condenses into a highly ordered structure called a chromosome. This process, called chromatin condenzation, is mediated by condensin.
Kinetochore	The kinetochore is the protein structure in eukaryotes which assembles on the centromere and links the chromosome to microtubule polymers from the mitotic spindle during mitosis.
Metaphase	The second stage of mitosis. During metaphase, all the cell's duplicated chromosomes are lined up at an imaginary plane equidistant between the poles of the mitotic spindle.
Anaphase	Anaphase is the stage of meiosis or mitosis when chromosomes separate in a eukaryotic cell. Each chromatid moves to opposite poles of the cell, the opposite ends of the mitotic spindle, near the microtubule organizing centers.
Microfilament	Microfilament refers to the thinnest of the three main kinds of protein fibers making up the cytoskeleton of a eukaryotic cell; a solid, helical rod composed of the globular protein actin.
Ribosome	A ribosome is an organelle composed of rRNA and ribosomal proteins. It translates mRNA into a polypeptide chain (e.g., a protein). It can be thought of as a factory that builds a protein from a set of genetic instructions.
Unicellular	Unicellular organisms carry out all the functions of life. Unicellular species are those whose members consist of a single cell throughout their life cycle. This latter qualification is significant since most multicellular organisms consist of a single cell at the beginning of their life cycles.
Sexual reproduction	The propagation of organisms involving the union of gametes from two parents is sexual reproduction.
Egg	An egg is the zygote, resulting from fertilization of the ovum. It nourishes and protects the embryo.
Haploid	Haploid cells bear one copy of each chromosome.
Zygote	Diploid cell formed by the union of sperm and egg is referred to as zygote.
Meiosis	In biology, meiosis is the process that transforms one diploid cell into four haploid cells in eukaryotes in order to redistribute the diploid's cell's genome. Meiosis forms the basis

Go to **Cram101.com** for the Practice Tests for this Chapter.

of sexual reproduction and can only occur in eukaryotes.

Homologous chromosome	Homologous chromosome refers to similarly constructed chromosomes with the same shape and that contain genes for the same traits.
Homologous	Homologous refers to describes organs or molecules that are similar because of their common evolutionary origin. Specifically it describes similarities in protein or nucleic acid sequence.
Chiasma	Chiasma refers to the microscopically visible site where crossing over has occurred between chromatids of homologous chromosomes during prophase I of meiosis.
Interkinesis	Period of time between meiosis I and meiosis II during which no DNA replication takes place is called interkinesis.
Species	Group of similarly constructed organisms capable of interbreeding and producing fertile offspring is a species.
Nondisjunction	Nondisjunction refers to an accident of meiosis or mitosis in which a pair of homologous chromosomes or a pair of sister chromatids fail to separate at anaphase.
Translocation	A chromosomal mutation in which a portion of one chromosome breaks off and becomes attached to another chromosome is referred to as translocation.
Diploid	Diploid cells have two copies (homologs) of each chromosome (both sex- and non-sex determining chromosomes), usually one from the mother and one from the father. Most somatic cells (body cells) of complex organisms are diploid.
Fetus	Fetus refers to a developing human from the ninth week of gestation until birth; has all the major structures of an adult.
Apoptosis	In biology, apoptosis is one of the main types of programmed cell death (PCD). As such, it is a process of deliberate life relinquishment by an unwanted cell in a multicellular organism.

Hemophilia	Hemophilia is the name of any of several hereditary genetic illnesses that impair the body's ability to control bleeding. Genetic deficiencies cause lowered plasma clotting factor activity so as to compromise blood-clotting; when a blood vessel is injured, a scab will not form and the vessel can continue to bleed excessively for a very long period of time.
Reproduction	Biological reproduction is the biological process by which new individual organisms are produced. Reproduction is a fundamental feature of all known life; each individual organism exists as the result of reproduction by an antecedent.
Species	Group of similarly constructed organisms capable of interbreeding and producing fertile offspring is a species.
Element	A chemical element, often called simply element, is a chemical substance that cannot be divided or changed into other chemical substances by any ordinary chemical technique. An element is a class of substances that contain the same number of protons in all its atoms.
Flower	A flower is the reproductive structure of a flowering plant. The flower structure contains the plant's reproductive organs, and its function is to produce seeds through sexual reproduction.
True-breeding	True-breeding refers to organisms for which sexual reproduction produces offspring with inherited trait identical to those of the parents. The organisms are homozygous for the characteristic under consideration.
Pollen	The male gametophyte in gymnosperms and angiosperms is referred to as pollen.
Gamete	A gamete is a specialized germ cell that unites with another gamete during fertilization in organisms that reproduce sexually. They are haploid cells; that is, they contain one complete set of chromosomes. When they unite they form a zygote—a cell having two complete sets of chromosomes and therefore diploid.
Population	Group of organisms of the same species occupying a certain area and sharing a common gene pool is referred to as population.
Genotype	The genotype is the specific genetic makeup (the specific genome) of an individual, usually in the form of DNA. It codes for the phenotype of that individual.
Phenotype	The phenotype of an individual organism is either its total physical appearance and constitution or a specific manifestation of a trait, such as size or eye color, that varies between individuals. It is determined to some extent by genotype.
Pollination	In seed plants, the delivery of pollen to the vicinity of the egg-producing megagametophyte is pollination.
Test cross	Test cross refers to a cross between an organism whose genotype for a certain trait is unknown and an organism that is homozygous recessive for that trait so the unknown genotype can be determined from that of the offspring. A procedure Mendel used to further test his hypotheses.
Homozygous	When an organism is referred to as being homozygous for a specific gene, it means that it carries two identical copies of that gene for a given trait on the two corresponding chromosomes.
Heterozygous	Heterozygous means that the organism carries a different version of that gene on each of the two corresponding chromosomes.
Genetics	Genetics is the science of genes, heredity, and the variation of organisms.
Joint	A joint (articulation) is the location at which two bones make contact (articulate). They are constructed to both allow movement and provide mechanical support.

Go to **Cram101.com** for the Practice Tests for this Chapter.

Homozygote	Homozygote refers to a diploid individual whose two copies of a gene are the same. An individual carrying identical alleles on both homologous chromosomes is said to be homozygous for that gene.
Heterozygote	Heterozygote refers to a diploid or polypoid individual carrying two different alleles of a gene on its two homologous chromosomes.
Allele	An allele is any one of a number of viable DNA codings of the same gene (sometimes the term refers to a non-gene sequence) occupying a given locus (position) on a chromosome.
Recessive allele	Recessive allele causes a phenotype (visible or detectable characteristic) that is only seen in a homozygous genotype (an organism that has two copies of the same allele). Every person has two copies of every gene, one from mother and one from father. If a genetic trait is recessive, a person only needs to inherit two copies of the gene for the trait to be expressed.
Pedigree	A record of one's ancestors, offspring, siblings, and their offspring that may be used to determine the pattern of certain genes or disease inheritance within a family is a pedigree.
Gene	Gene refers to a discrete unit of hereditary information consisting of a specific nucleotide sequence in DNA . Most of the genes of a eukaryote are located in its chromosomal DNA; a few are carried by the DNA of mitochondria and chloroplasts.
Mutation	Mutation refers to a change in the nucleotide sequence of DNA; the ultimate source of genetic diversity.
Monohybrid cross	A monohybrid cross, in genetics, is the mating between two heterozygous individuals. Generally, dominant characteristics are represented with a capital letter, A, and recessive characteristics are represented by a lower case letter, a.
Protein	A protein is a complex, high-molecular-weight organic compound that consists of amino acids joined by peptide bonds. They are essential to the structure and function of all living cells and viruses. Many are enzymes or subunits of enzymes.
Dominant allele	Dominant allele refers to an allele that exerts its phenotypic effect in the heterozygote.
Blood	Blood is a circulating tissue composed of fluid plasma and cells. The main function of blood is to supply nutrients (oxygen, glucose) and constitutional elements to tissues and to remove waste products.
Chromosome	A chromosome is, minimally, a very long, continuous piece of DNA, which contains many genes, regulatory elements and other intervening nucleotide sequences.
Hybrid	Hybrid refers to the offspring of parents of two different species or of two different varieties of one species; the offspring of two parents that differ in one or more inherited traits; an individual that is heterozygous for one or more pair of genes.
Hybridization	In molecular biology hybridization is the process of joining two complementary strands of
Pigment	Pigment is any material resulting in color in plant or animal cells which is the result of selective absorption.
Expressivity	Expressivity refers to variations of a phenotype in genetics. The term is used to qualitatively characterize the variance or extent of the phenotype.
Cell	The cell is the structural and functional unit of all living organisms, and is sometimes called the "building block of life."
Linkage	Linkage refers to the patterns of assortment of genes that are located on the same chromosome. Important because if the genes are located relatively far apart, crossing over is more likely to occur between them than if they are close together.

Chiasma	Chiasma refers to the microscopically visible site where crossing over has occurred between chromatids of homologous chromosomes during prophase I of meiosis.
Crossing over	An essential element of meiosis occurring during prophase when nonsister chromatids exchange portions of DNA strands is called crossing over.
Tissue	Group of similar cells which perform a common function is called tissue.
Dioecious	A plant having unisexual flowers, conifer cones, or functionally equivalent structures occurring on different individuals is called dioecious.
Egg	An egg is the zygote, resulting from fertilization of the ovum. It nourishes and protects the embryo.
Autosome	An autosome is a non-sex chromosome. It is an ordinary paired chromosome that is the same in both sexes of a species.
Sperm	Sperm refers to the male sex cell with three distinct parts at maturity: head, middle piece, and tail.
Sex chromosome	The X or Y chromosome in human beings that determines the sex of an individual. Females have two X chromosomes in diploid cells; males have an X and a Y chromosome. The sex chromosome comprises the 23rd chromosome pair in a karyotype.
Turner syndrome	Turner syndrome is a human genetic abnormality, caused by a nondisjunction in the sex chromosomes that occurs in females (1 out of every 2,500 births). Instead of the normal XX sex chromosomes, only one X chromosome is present and fully functional.
Y chromosome	Male sex chromosome that carries genes involved in sex determination is referred to as the Y chromosome. It contains the genes that cause testis development, thus determining maleness.
X chromosome	The X chromosome is the female sex chromosome that carries genes involved in sex determination. Females have two X chromosomes, while males have one X and one Y chromosome.
X-linked	X-linked refers to allele located on an X chromosome but may control a trait that has nothing to do with the sex characteristics of an individual.
Nucleus	In cell biology, the nucleus is found in all eukaryotic cells that contains most of the cell's genetic material. The nucleus has two primary functions: to control chemical reactions within the cytoplasm and to store information needed for cellular division.
Muscle	Muscle is a contractile form of tissue. It is one of the four major tissue types, the other three being epithelium, connective tissue and nervous tissue. Muscle contraction is used to move parts of the body, as well as to move substances within the body.

Go to **Cram101.com** for the Practice Tests for this Chapter.

Somatic cell	A somatic cell is generally taken to mean any cell forming the body of an organism.
Somatic	The term somatic refers to the body. It also refers to the part of the nervous system that controls voluntary movement and senzation and judges relative effort and weight, called proprioception.
Cell	The cell is the structural and functional unit of all living organisms, and is sometimes called the "building block of life."
Protein	A protein is a complex, high-molecular-weight organic compound that consists of amino acids joined by peptide bonds. They are essential to the structure and function of all living cells and viruses. Many are enzymes or subunits of enzymes.
Enzyme	An enzyme is a protein that catalyzes, or speeds up, a chemical reaction. They are essential to sustain life because most chemical reactions in biological cells would occur too slowly, or would lead to different products, without them.
Biology	Biology is the branch of science dealing with the study of life. It is concerned with the characteristics, classification, and behaviors of organisms, how species come into existence, and the interactions they have with each other and with the environment.
Population	Group of organisms of the same species occupying a certain area and sharing a common gene pool is referred to as population.
Bacteria	The domain that contains procaryotic cells with primarily diacyl glycerol diesters in their membranes and with bacterial rRNA. Bacteria also is a general term for organisms that are composed of procaryotic cells and are not multicellular.
Polysaccharide	Polymer made from sugar monomers is a polysaccharide. They are relatively complex carbohydrates.
Vaccine	A harmless variant or derivative of a pathogen used to stimulate a host organism's immune system to mount a long-term defense against the pathogen is referred to as vaccine.
Transformation	Transformation is the genetic alteration of a cell resulting from the introduction, uptake and expression of foreign genetic material (DNA or RNA).
Genetics	Genetics is the science of genes, heredity, and the variation of organisms.
Virus	Obligate intracellular parasite of living cells consisting of an outer capsid and an inner core of nucleic acid is referred to as virus. The term virus usually refers to those particles that infect eukaryotes whilst the term bacteriophage or phage is used to describe those infecting prokaryotes.
Bacteriophage	Bacteriophage refers to a virus that infects bacteria; also called a phage.
Element	A chemical element, often called simply element, is a chemical substance that cannot be divided or changed into other chemical substances by any ordinary chemical technique. An element is a class of substances that contain the same number of protons in all its atoms.
Radioactive	A term used to describe the property of releasing energy or particles from an unstable atom is called radioactive.
Nucleotide	A nucleotide is a chemical compound that consists of a heterocyclic base, a sugar, and one or more phosphate groups. In the most common nucleotides the base is a derivative of purine or pyrimidine, and the sugar is pentose - deoxyribose or ribose. They are the structural units of RNA and DNA.
Sugar	A sugar is the simplest molecule that can be identified as a carbohydrate. These include monosaccharides and disaccharides, trisaccharides and the oligosaccharides. The term "glyco-" indicates the presence of a sugar in an otherwise non-carbohydrate substance.

Go to **Cram101.com** for the Practice Tests for this Chapter.
And, **NEVER** highlight a book again!

Adenine	Adenine is one of the two purine nucleobases used in forming nucleotides of the nucleic acids DNA and RNA. In DNA, adenine (A) binds to thymine (T) via two hydrogen bonds to assist in stabilizing the nucleic acid structures. In RNA, adenine binds to uracil (U).
Double helix	Double helix refers to the form of native DNA, referring to its two adjacent polynucleotide strands wound into a spiral shape.
Helix	A helix is a twisted shape like a spring, screw or a spiral staircase. They are important in biology, as DNA and many proteins have spiral substructures, known a alpha helix.
Sugar-phosphate backbone	Sugar-phosphate backbone refers to the alternating chain of sugar and phosphate to which DNA and RNA nitrogenous bases are attached.
Atom	An atom is the smallest possible particle of a chemical element that retains its chemical properties.
Hydroxyl group	The term hydroxyl group is used to describe the functional group -OH when it is a substituent in an organic compound.
Mutation	Mutation refers to a change in the nucleotide sequence of DNA; the ultimate source of genetic diversity.
Cell division	Cell division is the process by which a cell (called the parent cell) divides into two cells (called daughter cells). Cell division is usually a small segment of a larger cell cycle. In meiosis, however, a cell is permanently transformed and cannot divide again.
Complementary base pairing	Hydrogen bonding between particular bases is a complementary base pairing.
Gradient	Gradient refers to a difference in concentration, pressure, or electrical charge between two regions.
Ion	Ion refers to an atom or molecule that has gained or lost one or more electrons, thus acquiring an electrical charge.
DNA replication	DNA replication is the process of copying a double-stranded DNA strand in a cell, prior to cell division. The two resulting double strands are identical (if the replication went well), and each of them consists of one original and one newly synthesized strand.
Semiconserva-ive replication	Semiconservative replication in DNA replication would produce two copies that each contained one of the original strands and one entirely new strand.
DNA polymerase	A DNA polymerase is an enzyme that assists in DNA replication. Such enzymes catalyze the polymerization of deoxyribonucleotides alongside a DNA strand, which they "read" and use as a template. The newly polymerized molecule is complementary to the template strand and identical to the template's partner strand.
Polymerase	DNA or RNA enzymes that catalyze the synthesis of nucleic acids on preexisting nucleic acid templates, assembling RNA from ribonucleotides or DNA from deoxyribonucleotides is referred to as polymerase.
Chromosome	A chromosome is, minimally, a very long, continuous piece of DNA, which contains many genes, regulatory elements and other intervening nucleotide sequences.
Molecule	A molecule is the smallest particle of a pure chemical substance that still retains its chemical composition and properties.
Active site	The active site of an enzyme is the binding site where catalysis occurs. The structure and chemical properties of the active site allow the recognition and binding of the substrate.
Template strand	The strand of the DNA double helix from which RNA is transcribed is called a template strand.

Go to **Cram101.com** for the Practice Tests for this Chapter.

DNA helicase	An enzyme that catalyzes the unwinding of the DNA double helix during DNA replication is referred to as DNA helicase.
Helicase	Helicase refers to a class of enzymes vital to all living organisms. Its function is to temporarily separate the two strands of a DNA double helix so that DNA or RNA synthesis can take place.
Colon	The colon is the part of the intestine from the cecum to the rectum. Its primary purpose is to extract water from feces.
Electrophoresis	Electrophoresis is the movement of an electrically charged substance under the influence of an electric field. This movement is due to the Lorentz force, which may be related to fundamental electrical properties of the body under study and the ambient electrical conditions.
Polymerase chain reaction	Polymerase chain reaction is a molecular biology technique for enzymatically replicating DNA without using a living organism, such as E. coli or yeast. The technique allows a small amount of the DNA molecule to be amplified many times, in an exponential manner.
Denature	Denature is a structural change in biomolecules such as nucleic acids and proteins, such that they are no longer in their native state, and their shape which allows for optimal activity.
Denaturation	Denaturation is a structural change in biomolecules such as nucleic acids and proteins, such that they are no longer in their native state, and their shape which allows for optimal activity.

Go to **Cram101.com** for the Practice Tests for this Chapter.

Gene	Gene refers to a discrete unit of hereditary information consisting of a specific nucleotide sequence in DNA . Most of the genes of a eukaryote are located in its chromosomal DNA; a few are carried by the DNA of mitochondria and chloroplasts.
Mutagen	A chemical or physical agent that interacts with DNA and causes a mutation is referred to as mutagen.
Amino acid	An amino acid is any molecule that contains both amino and carboxylic acid functional groups. They are the basic structural building units of proteins. They form short polymer chains called peptides or polypeptides which in turn form structures called proteins.
Acid	An acid is a water-soluble, sour-tasting chemical compound that when dissolved in water, gives a solution with a pH of less than 7.
Metabolic pathway	A metabolic pathway is a series of chemical reactions occurring within a cell, catalyzed by enzymes, to achieve in either the formation of a metabolic product to be used or stored by the cell, or the initiation of another metabolic pathway.
Mutation	Mutation refers to a change in the nucleotide sequence of DNA; the ultimate source of genetic diversity.
Transcription	Transcription is the process through which a DNA sequence is enzymatically copied by an RNA polymerase to produce a complementary RNA. Or, in other words, the transfer of genetic information from DNA into RNA.
Sugar	A sugar is the simplest molecule that can be identified as a carbohydrate. These include monosaccharides and disaccharides, trisaccharides and the oligosaccharides. The term "glyco-" indicates the presence of a sugar in an otherwise non-carbohydrate substance.
Virus	Obligate intracellular parasite of living cells consisting of an outer capsid and an inner core of nucleic acid is referred to as virus. The term virus usually refers to those particles that infect eukaryotes whilst the term bacteriophage or phage is used to describe those infecting prokaryotes.
Nucleotide	A nucleotide is a chemical compound that consists of a heterocyclic base, a sugar, and one or more phosphate groups. In the most common nucleotides the base is a derivative of purine or pyrimidine, and the sugar is pentose - deoxyribose or ribose. They are the structural units of RNA and DNA.
Tumor	An abnormal mass of cells that forms within otherwise normal tissue is a tumor. This growth can be either malignant or benign
Host	Host is an organism that harbors a parasite, mutual partner, or commensal partner; or a cell infected by a virus.
Complementary DNA	Complementary DNA is DNA synthesized from a mature mRNA template. It is often used to clone eukaryotic genes in prokaryotes.
RNA polymerase	The enzyme RNA polymerase is a nucleotidyltransferase that polymerises ribonucleotides in accordance with the information present in DNA. RNA polymerase enzymes are essential and are found in all organisms.
Polymerase	DNA or RNA enzymes that catalyze the synthesis of nucleic acids on preexisting nucleic acid templates, assembling RNA from ribonucleotides or DNA from deoxyribonucleotides is referred to as polymerase.
Promoter	In genetics, a promoter is a DNA sequence that enables a gene to be transcribed. The promoter is recognized by RNA polymerase, which then initiates transcription.
Template strand	The strand of the DNA double helix from which RNA is transcribed is called a template strand.

Go to **Cram101.com** for the Practice Tests for this Chapter.

DNA polymerase	A DNA polymerase is an enzyme that assists in DNA replication. Such enzymes catalyze the polymerization of deoxyribonucleotides alongside a DNA strand, which they "read" and use as a template. The newly polymerized molecule is complementary to the template strand and identical to the template's partner strand.
DNA replication	DNA replication is the process of copying a double-stranded DNA strand in a cell, prior to cell division. The two resulting double strands are identical (if the replication went well), and each of them consists of one original and one newly synthesized strand.
Translation	Translation is the second process of protein biosynthesis. In translation, messenger RNA is decoded to produce a specific polypeptide according to the rules specified by the genetic code.
Genetic code	The genetic code is a set of rules that maps DNA sequences to proteins in the living cell, and is employed in the process of protein synthesis. Nearly all living things use the same genetic code, called the standard genetic code, although a few organisms use minor variations of the standard code.
Codon	Codon refers to a three-nucleotide sequence in mRNA that specifies a particular amino acid or polypeptide termination signal; the basic unit of the genetic code.
Start codon	Start codon refers to on mRNA, the specific threenucleotide sequence to which an initiator tRNA molecule binds, starting translation of genetic information.
Uracil	Uracil is one of the four RNA nucleobases, replacing thymine as found in DNA. Just like thymine, uracil can form a base pair with adenine via two hydrogen bonds, but it lacks the methyl group present in thymine. Uracil, in comparison to thymine, will more readily degenerate into cytosine.
Protein synthesis	The process whereby the tRNA utilizes the mRNA as a guide to arrange the amino acids in their proper sequence according to the genetic information in the chemical code of DNA is referred to as protein synthesis.
Protein	A protein is a complex, high-molecular-weight organic compound that consists of amino acids joined by peptide bonds. They are essential to the structure and function of all living cells and viruses. Many are enzymes or subunits of enzymes.
Ribosome	A ribosome is an organelle composed of rRNA and ribosomal proteins. It translates mRNA into a polypeptide chain (e.g., a protein). It can be thought of as a factory that builds a protein from a set of genetic instructions.
Molecule	A molecule is the smallest particle of a pure chemical substance that still retains its chemical composition and properties.
Anticodon	An anticodon is a unit made up of three nucleotides which play an important role in various DNA cycles, including RNA translation. Each tRNA contains a specific anticodon triplet sequence that can base-pair to one or more codons for an amino acid.
Species	Group of similarly constructed organisms capable of interbreeding and producing fertile offspring is a species.
Enzyme	An enzyme is a protein that catalyzes, or speeds up, a chemical reaction. They are essential to sustain life because most chemical reactions in biological cells would occur too slowly, or would lead to different products, without them.
Polypeptide	Polypeptide refers to polymer of many amino acids linked by peptide bonds.
Stop codon	In mRNA, one of three triplets that signal gene translation to stop is called stop codon.
Antibiotic	Antibiotic refers to substance such as penicillin or streptomycin that is toxic to microorganisms. Usually a product of a particular microorvanism or plant.

Go to **Cram101.com** for the Practice Tests for this Chapter.

Microorganism	A microorganism is an organism that is so small that it is microscopic (invisible to the naked eye). They are often illustrated using single-celled, or unicellular organisms; however, some unicellular protists are visible to the naked eye, and some multicellular species are microscopic.
Eukaryotic cell	Eukaryotic cell refers to a type of cell that has a membrane-enclosed nucleus and other membrane enclosed organelles. All organisms except bacteria and archaea are composed of eukaryotic cells.
Cell	The cell is the structural and functional unit of all living organisms, and is sometimes called the "building block of life."
Receptor protein	Protein located in the plasma membrane or within the cell that binds to a substance that alters some metabolic aspect of the cell is referred to as receptor protein. It will only link up with a substance that has a certain shape that allows it to bind to the receptor.
Receptor	A receptor is a protein on the cell membrane or within the cytoplasm or cell nucleus that binds to a specific molecule (a ligand), such as a neurotransmitter, hormone, or other substance, and initiates the cellular response to the ligand. Receptor, in immunology, the region of an antibody which shows recognition of an antigen.
Golgi apparatus	Golgi apparatus refers to an organelle in eukaryotic cells consisting of stacks of membranous sacs that modify, store, and ship products of the endoplasmic reticulum.
Golgi	Golgi discovered a method of staining nervous tissue which would stain a limited number of cells at random, in their entirety. This enabled him to view the paths of nerve cells in the brain for the first time. He called his discovery the black reaction. It is now known universally as the Golgi stain.
Protease	Protease refers to an enzyme that breaks peptide bonds between amino acids of proteins.
Unicellular	Unicellular organisms carry out all the functions of life. Unicellular species are those whose members consist of a single cell throughout their life cycle. This latter qualification is significant since most multicellular organisms consist of a single cell at the beginning of their life cycles.
Multicellular	Multicellular organisms are those organisms consisting of more than one cell, and having differentiated cells that perform specialized functions. Most life that can be seen with the naked eye is multicellular, as are all animals (i.e. members of the kingdom Animalia) and plants (i.e. members of the kingdom Plantae).
Gamete	A gamete is a specialized germ cell that unites with another gamete during fertilization in organisms that reproduce sexually. They are haploid cells; that is, they contain one complete set of chromosomes. When they unite they form a zygote—a cell having two complete sets of chromosomes and therefore diploid.
Phenotype	The phenotype of an individual organism is either its total physical appearance and constitution or a specific manifestation of a trait, such as size or eye color, that varies between individuals. It is determined to some extent by genotype.
Blood	Blood is a circulating tissue composed of fluid plasma and cells. The main function of blood is to supply nutrients (oxygen, glucose) and constitutional elements to tissues and to remove waste products.
Allele	An allele is any one of a number of viable DNA codings of the same gene (sometimes the term refers to a non-gene sequence) occupying a given locus (position) on a chromosome.
Bacterium	Most bacterium are microscopic and unicellular, with a relatively simple cell structure lacking a cell nucleus, and organelles such as mitochondria and chloroplasts. They are the most abundant of all organisms. They are ubiquitous in soil, water, and as symbionts of other

Go to **Cram101.com** for the Practice Tests for this Chapter.

organisms.

Chromosome	A chromosome is, minimally, a very long, continuous piece of DNA, which contains many genes, regulatory elements and other intervening nucleotide sequences.
Homozygous	When an organism is referred to as being homozygous for a specific gene, it means that it carries two identical copies of that gene for a given trait on the two corresponding chromosomes.
Evolution	In biology, evolution is the process by which novel traits arise in populations and are passed on from generation to generation. Its action over large stretches of time explains the origin of new species and ultimately the vast diversity of the biological world.
Deletion	Deletion refers to the loss of one or more nucleotides from a gene by mutation; the loss of a fragment of a chromosome.
Homologous chromosome	Homologous chromosome refers to similarly constructed chromosomes with the same shape and that contain genes for the same traits.
Homologous	Homologous refers to describes organs or molecules that are similar because of their common evolutionary origin. Specifically it describes similarities in protein or nucleic acid sequence.
Inversion	Inversion refers to a change in a chromosome resulting from reattachment in a reverse direction of a chromosome fragment to the original chromosome. Mutagens and errors during meiosis can cause them.
Synapsis	Synapsis is the currently unexplained phenomenon of two homologous chromosomes coming together and lining up side-by-side resulting in a tetrad, or two homologous chromosomes that stay in close association during the first two phases of meiosis I.
Genome	The genome of an organism is the whole hereditary information of an organism that is encoded in the DNA (or, for some viruses, RNA). This includes both the genes and the non-coding sequences. The genome of an organism is a complete DNA sequence of one set of chromosomes.
Radiation	The emission of electromagnetic waves by all objects warmer than absolute zero is referred to as radiation.

Bacterium	Most bacterium are microscopic and unicellular, with a relatively simple cell structure lacking a cell nucleus, and organelles such as mitochondria and chloroplasts. They are the most abundant of all organisms. They are ubiquitous in soil, water, and as symbionts of other organisms.
Haploid	Haploid cells bear one copy of each chromosome.
Bacteria	The domain that contains procaryotic cells with primarily diacyl glycerol diesters in their membranes and with bacterial rRNA. Bacteria also is a general term for organisms that are composed of procaryotic cells and are not multicellular.
Biotechnology	Biotechnology refers to the use of living organisms to perform useful tasks; today, usually involves DNA technology.
Nucleic acid	A nucleic acid is a complex, high-molecular-weight biochemical macromolecule composed of nucleotide chains that convey genetic information. The most common are deoxyribonucleic acid (DNA) and ribonucleic acid (RNA). They are found in all living cells and viruses.
Acid	An acid is a water-soluble, sour-tasting chemical compound that when dissolved in water, gives a solution with a pH of less than 7.
Host	Host is an organism that harbors a parasite, mutual partner, or commensal partner; or a cell infected by a virus.
Cell	The cell is the structural and functional unit of all living organisms, and is sometimes called the "building block of life."
Obligate intracellular parasites	Obligate intracellular parasites are microorganisms that cannot reproduce outside their host cell, they force or compel that host to assist in propagating themselves.
Parasite	A parasite is an organism that spends a significant portion of its life in or on the living tissue of a host organism and which causes harm to the host without immediately killing it. They also commonly show highly specialized adaptations allowing them to exploit host resources.
Lipid	Lipid is one class of aliphatic hydrocarbon-containing organic compounds essential for the structure and function of living cells. They are characterized by being water-insoluble but soluble in nonpolar organic solvents.
Virus	Obligate intracellular parasite of living cells consisting of an outer capsid and an inner core of nucleic acid is referred to as virus. The term virus usually refers to those particles that infect eukaryotes whilst the term bacteriophage or phage is used to describe those infecting prokaryotes.
Receptor	A receptor is a protein on the cell membrane or within the cytoplasm or cell nucleus that binds to a specific molecule (a ligand), such as a neurotransmitter, hormone, or other substance, and initiates the cellular response to the ligand. Receptor, in immunology, the region of an antibody which shows recognition of an antigen.
Cell wall	Cell wall refers to a protective layer external to the plasma membrane in plant cells, bacteria, fungi, and some protists; protects the cell and helps maintain its shape.
Enzyme	An enzyme is a protein that catalyzes, or speeds up, a chemical reaction. They are essential to sustain life because most chemical reactions in biological cells would occur too slowly, or would lead to different products, without them.
Gene	Gene refers to a discrete unit of hereditary information consisting of a specific nucleotide sequence in DNA . Most of the genes of a eukaryote are located in its chromosomal DNA; a few are carried by the DNA of mitochondria and chloroplasts.

Go to **Cram101.com** for the Practice Tests for this Chapter.

Genome	The genome of an organism is the whole hereditary information of an organism that is encoded in the DNA (or, for some viruses, RNA). This includes both the genes and the non-coding sequences. The genome of an organism is a complete DNA sequence of one set of chromosomes.
Vertebrate	Vertebrate is a subphylum of chordates, specifically, those with backbones or spinal columns. They started to evolve about 530 million years ago during the Cambrian explosion, which is part of the Cambrian period.
Arthropod	Arthropod refers to member of a phylum of invertebrates that contains among other groups crustaceans and insects that have an exoskeleton and jointed appendages.
Endocytosis	Endocytosis is a process where cells absorb material (molecules or other cells) from outside by engulfing it with their cell membranes.
Vesicle	In cell biology, a vesicle is a relatively small and enclosed compartment, separated from the cytosol by at least one lipid bilayer.
Complementary base pairing	Hydrogen bonding between particular bases is a complementary base pairing.
Insect	An arthropod that usually has three body segments , three pairs of legs, and one or two pairs of wings is called an insect. They are the largest and (on land) most widely-distributed taxon within the phylum Arthropoda. They comprise the most diverse group of animals on the earth, with around 925,000 species described
Sexual reproduction	The propagation of organisms involving the union of gametes from two parents is sexual reproduction.
Reproduction	Biological reproduction is the biological process by which new individual organisms are produced. Reproduction is a fundamental feature of all known life; each individual organism exists as the result of reproduction by an antecedent.
Organ	Organ refers to a structure consisting of several tissues adapted as a group to perform specific functions.
Genetic recombination	Genetic recombination refers to the production, by crossing over and/or independent assortment of chromosomes during meiosis, of offspring with allele combinations different from those in the parents.
Recombination	Genetic recombination is the transmission-genetic process by which the combinations of alleles observed at different loci in two parental individuals become shuffled in offspring individuals.
Mutation	Mutation refers to a change in the nucleotide sequence of DNA; the ultimate source of genetic diversity.
Genotype	The genotype is the specific genetic makeup (the specific genome) of an individual, usually in the form of DNA. It codes for the phenotype of that individual.
Conjugation	The direct transfer of genetic material from one cell to another while the cells are temporarily joined is called conjugation.
Chromosome	A chromosome is, minimally, a very long, continuous piece of DNA, which contains many genes, regulatory elements and other intervening nucleotide sequences.
Antibiotic	Antibiotic refers to substance such as penicillin or streptomycin that is toxic to microorganisms. Usually a product of a particular microorvanism or plant.
Element	A chemical element, often called simply element, is a chemical substance that cannot be divided or changed into other chemical substances by any ordinary chemical technique. An element is a class of substances that contain the same number of protons in all its atoms.

Species	Group of similarly constructed organisms capable of interbreeding and producing fertile offspring is a species.
Protein	A protein is a complex, high-molecular-weight organic compound that consists of amino acids joined by peptide bonds. They are essential to the structure and function of all living cells and viruses. Many are enzymes or subunits of enzymes.
Ammonia	Ammonia is a compound of nitrogen and hydrogen with the formula NH_3. At standard temperature and pressure ammonia is a gas. It is toxic and corrosive to some materials, and has a characteristic pungent odor.
Transcription	Transcription is the process through which a DNA sequence is enzymatically copied by an RNA polymerase to produce a complementary RNA. Or, in other words, the transfer of genetic information from DNA into RNA.
Protein synthesis	The process whereby the tRNA utilizes the mRNA as a guide to arrange the amino acids in their proper sequence according to the genetic information in the chemical code of DNA is referred to as protein synthesis.
Metabolism	Metabolism is the biochemical modification of chemical compounds in living organisms and cells. This includes the biosynthesis of complex organic molecules (anabolism) and their breakdown (catabolism).
Repressor	A repressor is a DNA-binding protein that regulates the expression of one or more genes by decreasing the rate of transcription.
Operon	An operon is a group of key nucleotide sequences including an operator, a common promoter, and one or more structural genes that are controlled as a unit to produce messenger RNA (mRNA).
Promoter	In genetics, a promoter is a DNA sequence that enables a gene to be transcribed. The promoter is recognized by RNA polymerase, which then initiates transcription.
Lactose	Lactose is a disaccharide that makes up around 2-8% of the solids in milk. Lactose is a disaccharide consisting of two subunits, a galactose and a glucose linked together.
Amino acid	An amino acid is any molecule that contains both amino and carboxylic acid functional groups. They are the basic structural building units of proteins. They form short polymer chains called peptides or polypeptides which in turn form structures called proteins.
Glucose	Glucose, a simple monosaccharide sugar, is one of the most important carbohydrates and is used as a source of energy in animals and plants. Glucose is one of the main products of photosynthesis and starts respiration.
Gene expression	Gene expression is the process by which a gene's information is converted into the structures and functions of a cell. Gene expression is a multi-step process that begins with transcription, post transcriptional modification and translation, followed by folding, post-translational modification and targeting.
DNA sequencing	The process of determining the chemical composition of a DNA molecule is called DNA sequencing.
Sequencing	Determining the order of nucleotides in a DNA or RNA molecule or the order of amino acids in a protein is sequencing.
Bacteriophage	Bacteriophage refers to a virus that infects bacteria; also called a phage.
Hybridization	In molecular biology hybridization is the process of joining two complementary strands of
Transposon	DNA segment that carries the genes required for transposition and moves about the chromosome; if it contains genes other than those required for transposition, it may be called a

Go to **Cram101.com** for the Practice Tests for this Chapter.

composite transposon. Often the name is reserved only for transposable elements that also contain genes unrelated to transposition.

Virus	Obligate intracellular parasite of living cells consisting of an outer capsid and an inner core of nucleic acid is referred to as virus. The term virus usually refers to those particles that infect eukaryotes whilst the term bacteriophage or phage is used to describe those infecting prokaryotes.
Prokaryote	A prokaryote is an organisms without a cell nucleus, or indeed any other membrane-bound organelles, in most cases unicellular (in rare cases, multicellular). This is in contrast to a eukaryote, organisms that have cell nuclei and may be variously unicellular or multicellular.
Cell	The cell is the structural and functional unit of all living organisms, and is sometimes called the "building block of life."
Chromosome	A chromosome is, minimally, a very long, continuous piece of DNA, which contains many genes, regulatory elements and other intervening nucleotide sequences.
Nuclear envelope	The nuclear envelope refers to the double membrane of the nucleus that encloses genetic material in eukaryotic cells. It separates the contents of the nucleus (DNA in particular) from the cytosol.
Nucleus	In cell biology, the nucleus is found in all eukaryotic cells that contains most of the cell's genetic material. The nucleus has two primary functions: to control chemical reactions within the cytoplasm and to store information needed for cellular division.
Genome	The genome of an organism is the whole hereditary information of an organism that is encoded in the DNA (or, for some viruses, RNA). This includes both the genes and the non-coding sequences. The genome of an organism is a complete DNA sequence of one set of chromosomes.
Gene	Gene refers to a discrete unit of hereditary information consisting of a specific nucleotide sequence in DNA . Most of the genes of a eukaryote are located in its chromosomal DNA; a few are carried by the DNA of mitochondria and chloroplasts.
Nervous system	The nervous system of an animal coordinates the activity of the muscles, monitors the organs, constructs and processes input from the senses, and initiates actions.
Multicellular	Multicellular organisms are those organisms consisting of more than one cell, and having differentiated cells that perform specialized functions. Most life that can be seen with the naked eye is multicellular, as are all animals (i.e. members of the kingdom Animalia) and plants (i.e. members of the kingdom Plantae).
Fruit	A fruit is the ripened ovary—together with seeds—of a flowering plant. In many species, the fruit incorporates the ripened ovary and surrounding tissues.
Human genome	The human genome is the genome of Homo sapiens. It is made up of 23 chromosome pairs with a total of about 3 billion DNA base pairs.
DNA polymerase	A DNA polymerase is an enzyme that assists in DNA replication. Such enzymes catalyze the polymerization of deoxyribonucleotides alongside a DNA strand, which they "read" and use as a template. The newly polymerized molecule is complementary to the template strand and identical to the template's partner strand.
Polymerase	DNA or RNA enzymes that catalyze the synthesis of nucleic acids on preexisting nucleic acid templates, assembling RNA from ribonucleotides or DNA from deoxyribonucleotides is referred to as polymerase.
Bone marrow	Bone marrow is the tissue comprising the center of large bones. It is the place where new blood cells are produced. Bone marrow contains two types of stem cells: hemopoietic (which can produce blood cells) and stromal (which can produce fat, cartilage and bone).
Enzyme	An enzyme is a protein that catalyzes, or speeds up, a chemical reaction. They are essential

Go to **Cram101.com** for the Practice Tests for this Chapter.

to sustain life because most chemical reactions in biological cells would occur too slowly, or would lead to different products, without them.

Cancer	Cancer is a class of diseases or disorders characterized by uncontrolled division of cells and the ability of these cells to invade other tissues, either by direct growth into adjacent tissue through invasion or by implantation into distant sites by metastasis.
Repetitive DNA	Repetitive DNA refers to nucleotide sequences that are present in many copies in the DNA of a genome. The repeated sequences may be long or short and may be located next to each other or dispersed in the DNA.
Protein	A protein is a complex, high-molecular-weight organic compound that consists of amino acids joined by peptide bonds. They are essential to the structure and function of all living cells and viruses. Many are enzymes or subunits of enzymes.
Transposon	DNA segment that carries the genes required for transposition and moves about the chromosome; if it contains genes other than those required for transposition, it may be called a composite transposon. Often the name is reserved only for transposable elements that also contain genes unrelated to transposition.
Insertion	At DNA level, an insertion means the insertion of a few base pairs into a genetic sequence. This can often happen in microsatellite regions due to the DNA polymerase slipping.
Element	A chemical element, often called simply element, is a chemical substance that cannot be divided or changed into other chemical substances by any ordinary chemical technique. An element is a class of substances that contain the same number of protons in all its atoms.
Organelle	In cell biology, an organelle is one of several structures with specialized functions, suspended in the cytoplasm of a eukaryotic cell.
Promoter	In genetics, a promoter is a DNA sequence that enables a gene to be transcribed. The promoter is recognized by RNA polymerase, which then initiates transcription.
Vertebrate	Vertebrate is a subphylum of chordates, specifically, those with backbones or spinal columns. They started to evolve about 530 million years ago during the Cambrian explosion, which is part of the Cambrian period.
Evolution	In biology, evolution is the process by which novel traits arise in populations and are passed on from generation to generation. Its action over large stretches of time explains the origin of new species and ultimately the vast diversity of the biological world.
Hemoglobin	Hemoglobin is the iron-containing oxygen-transport metalloprotein in the red cells of the blood in mammals and other animals. Hemoglobin transports oxygen from the lungs to the rest of the body, such as to the muscles, where it releases the oxygen load.
Molecule	A molecule is the smallest particle of a pure chemical substance that still retains its chemical composition and properties.
Ribosome	A ribosome is an organelle composed of rRNA and ribosomal proteins. It translates mRNA into a polypeptide chain (e.g., a protein). It can be thought of as a factory that builds a protein from a set of genetic instructions.
Codon	Codon refers to a three-nucleotide sequence in mRNA that specifies a particular amino acid or polypeptide termination signal; the basic unit of the genetic code.
Intron	An intron is a section of DNA within a gene that does not encode part of the protein that the gene produces, and is spliced out of the mRNA that is transcribed from the gene before it is translated.
Receptor	A receptor is a protein on the cell membrane or within the cytoplasm or cell nucleus that binds to a specific molecule (a ligand), such as a neurotransmitter, hormone, or other

Go to **Cram101.com** for the Practice Tests for this Chapter.

	substance, and initiates the cellular response to the ligand. Receptor, in immunology, the region of an antibody which shows recognition of an antigen.
Gene expression	Gene expression is the process by which a gene's information is converted into the structures and functions of a cell. Gene expression is a multi-step process that begins with transcription, post transcriptional modification and translation, followed by folding, post-translational modification and targeting.
Transcription	Transcription is the process through which a DNA sequence is enzymatically copied by an RNA polymerase to produce a complementary RNA. Or, in other words, the transfer of genetic information from DNA into RNA.
Cytoplasm	Cytoplasm refers to everything inside a cell between the plasma membrane and the nucleus; consists of a semifluid medium and organelles.
Translation	Translation is the second process of protein biosynthesis. In translation, messenger RNA is decoded to produce a specific polypeptide according to the rules specified by the genetic code.
Brain	The part of the central nervous system involved in regulating and controlling body activity and interpreting information from the senses transmitted through the nervous system is referred to as the brain.
RNA polymerase	The enzyme RNA polymerase is a nucleotidyltransferase that polymerises ribonucleotides in accordance with the information present in DNA. RNA polymerase enzymes are essential and are found in all organisms.
Regulator protein	A protein that influences the activities that occur in an organism, for example, enzymes and some hormones is called a regulator protein.
Enhancer	An enhancer is a short region of DNA that can be bound with proteins to enhance transcription levels of genes in a gene-cluster.
Operon	An operon is a group of key nucleotide sequences including an operator, a common promoter, and one or more structural genes that are controlled as a unit to produce messenger RNA (mRNA).
Salt	Salt is a term used for ionic compounds composed of positively charged cations and negatively charged anions, so that the product is neutral and without a net charge.
Helix	A helix is a twisted shape like a spring, screw or a spiral staircase. They are important in biology, as DNA and many proteins have spiral substructures, known a alpha helix.
Hydrophobic	Hydrophobic refers to being electrically neutral and nonpolar, and thus prefering other neutral and nonpolar solvents or molecular environments. Hydrophobic is often used interchangeably with "oily" or "lipophilic."
Nucleosome	Nucleosome refers to the beadlike unit of DNA packing in a eukaryotic cell; consists of DNA wound around a protein core made up of eight histone molecules.
Chromatin	Chromatin refers to the combination of DNA and proteins that constitute chromosomes; often used to refer to the diffuse, very extended form taken by the chromosomes when a eukaryotic cell is not dividing.
Mammal	Homeothermic vertebrate characterized especially by the presence of hair and mammary glands is a mammal.
Barr body	In those species in which sex is determined by the presence of the Y or W chromosome rather than the diploidy of the X or Z, a Barr body is the inactive X chromosome in a female cell, or the inactive Z in a male, rendered inactive in a process called Lyonization.

Go to **Cram101.com** for the Practice Tests for this Chapter.

Cytosine	Cytosine is one of the 5 main nucleobases used in storing and transporting genetic information within a cell in the nucleic acids DNA and RNA. It is a pyrimidine derivative, with a heterocyclic aromatic ring and two substituents attached. The nucleoside of cytosine is cytidine.
X chromosome	The X chromosome is the female sex chromosome that carries genes involved in sex determination. Females have two X chromosomes, while males have one X and one Y chromosome.
Cell division	Cell division is the process by which a cell (called the parent cell) divides into two cells (called daughter cells). Cell division is usually a small segment of a larger cell cycle. In meiosis, however, a cell is permanently transformed and cannot divide again.
Repressor	A repressor is a DNA-binding protein that regulates the expression of one or more genes by decreasing the rate of transcription.
Allele	An allele is any one of a number of viable DNA codings of the same gene (sometimes the term refers to a non-gene sequence) occupying a given locus (position) on a chromosome.
Exon	An exon is the region of DNA within a gene that is not spliced out from the transcribed RNA and is retained in the final messenger RNA (mRNA) molecule.
Cell cycle	An orderly sequence of events that extends from the time a eukaryotic cell divides to form two daughter cells to the time those daughter cells divide again is called cell cycle.
Target cell	A cell that responds to a regulatory signal, such as a hormone is a target cell.
Oocyte	An oocyte is a female gametocyte. Such that an oocyte is large and essentially stationary. The oocyte becomes functional when a lala (male gametocyte) attaches to it, thus allowing the meiosis of the secondary oocyte to occur.
Fertilization	Fertilization is fusion of gametes to form a new organism. In animals, the process involves a sperm fusing with an ovum, which eventually leads to the development of an embryo.
Ion	Ion refers to an atom or molecule that has gained or lost one or more electrons, thus acquiring an electrical charge.
Polypeptide	Polypeptide refers to polymer of many amino acids linked by peptide bonds.
Amino acid	An amino acid is any molecule that contains both amino and carboxylic acid functional groups. They are the basic structural building units of proteins. They form short polymer chains called peptides or polypeptides which in turn form structures called proteins.
Acid	An acid is a water-soluble, sour-tasting chemical compound that when dissolved in water, gives a solution with a pH of less than 7.

Cell	The cell is the structural and functional unit of all living organisms, and is sometimes called the "building block of life."
Brain	The part of the central nervous system involved in regulating and controlling body activity and interpreting information from the senses transmitted through the nervous system is referred to as the brain.
Bacteria	The domain that contains procaryotic cells with primarily diacyl glycerol diesters in their membranes and with bacterial rRNA. Bacteria also is a general term for organisms that are composed of procaryotic cells and are not multicellular.
Circulatory system	The circulatory system or cardiovascular system is the organ system which circulates blood around the body of most animals.
Solute	Substance that is dissolved in a solvent, forming a solution is referred to as a solute.
Receptor	A receptor is a protein on the cell membrane or within the cytoplasm or cell nucleus that binds to a specific molecule (a ligand), such as a neurotransmitter, hormone, or other substance, and initiates the cellular response to the ligand. Receptor, in immunology, the region of an antibody which shows recognition of an antigen.
Protein	A protein is a complex, high-molecular-weight organic compound that consists of amino acids joined by peptide bonds. They are essential to the structure and function of all living cells and viruses. Many are enzymes or subunits of enzymes.
Promoter	In genetics, a promoter is a DNA sequence that enables a gene to be transcribed. The promoter is recognized by RNA polymerase, which then initiates transcription.
Transduction	In physiology, transduction is transportation of a stimuli to the nervous system. In genetics, transduction is the transfer of viral, bacterial, or both bacterial and viral DNA from one cell to another via bacteriophage.
Receptor protein	Protein located in the plasma membrane or within the cell that binds to a substance that alters some metabolic aspect of the cell is referred to as receptor protein. It will only link up with a substance that has a certain shape that allows it to bind to the receptor.
Molecule	A molecule is the smallest particle of a pure chemical substance that still retains its chemical composition and properties.
Enzyme	An enzyme is a protein that catalyzes, or speeds up, a chemical reaction. They are essential to sustain life because most chemical reactions in biological cells would occur too slowly, or would lead to different products, without them.
Plasma membrane	Membrane surrounding the cytoplasm that consists of a phospholipid bilayer with embedded proteins is referred to as plasma membrane.
Plasma	In physics and chemistry, a plasma is an ionized gas, and is usually considered to be a distinct phase of matter. "Ionized" in this case means that at least one electron has been dissociated from a significant fraction of the molecules.
Steroid	A steroid is a lipid characterized by a carbon skeleton with four fused rings. Different steroids vary in the functional groups attached to these rings. Hundreds of distinct steroids have been identified in plants and animals. Their most important role in most living systems is as hormones.
Channel protein	Membrane transport protein that forms an aqueous pore in the membrane through which a specific solute, usually an ion, can pass is the channel protein.
Acetylcholine	The chemical compound acetylcholine was the first neurotransmitter to be identified. It is a chemical transmitter in both the peripheral nervous system (PNS) and central nervous system (CNS) in many organisms including humans.

Insulin	Insulin is a polypeptide hormone that regulates carbohydrate metabolism. Apart from being the primary effector in carbohydrate homeostasis, it also has a substantial effect on small vessel muscle tone, controls storage and release of fat (triglycerides) and cellular uptake of both amino acids and some electrolytes.
Effector	An effector is a molecule (originally referring to small molecules but now encompassing any regulatory molecule, includes proteins) that binds to a protein and thereby alters the activity of that protein.
Epinephrine	Epinephrine is a hormone and a neurotransmitter. Epinephrine plays a central role in the short-term stress reaction—the physiological response to threatening or exciting conditions (fight-or-flight response). It is secreted by the adrenal medulla.
Blood vessel	A blood vessel is a part of the circulatory system and function to transport blood throughout the body. The most important types, arteries and veins, are so termed because they carry blood away from or towards the heart, respectively.
Blood	Blood is a circulating tissue composed of fluid plasma and cells. The main function of blood is to supply nutrients (oxygen, glucose) and constitutional elements to tissues and to remove waste products.
Lipid bilayer	A lipid bilayer is a membrane or zone of a membrane composed of lipid molecules (usually phospholipids). The lipid bilayer is a critical component of all biological membranes, including cell membranes, and is a prerequisite for cell-based organisms.
Lipid	Lipid is one class of aliphatic hydrocarbon-containing organic compounds essential for the structure and function of living cells. They are characterized by being water-insoluble but soluble in nonpolar organic solvents.
Cytoplasm	Cytoplasm refers to everything inside a cell between the plasma membrane and the nucleus; consists of a semifluid medium and organelles.
Hormone	A hormone is a chemical messenger from one cell to another. All multicellular organisms produce hormones. The best known hormones are those produced by endocrine glands of vertebrate animals, but hormones are produced by nearly every organ system and tissue type in a human or animal body. Hormone molecules are secreted directly into the bloodstream, they move by circulation or diffusion to their target cells, which may be nearby cells in the same tissue or cells of a distant organ of the body.
Skeletal muscle	Skeletal muscle is a type of striated muscle, attached to the skeleton. They are used to facilitate movement, by applying force to bones and joints; via contraction. They generally contract voluntarily (via nerve stimulation), although they can contract involuntarily.
Muscle	Muscle is a contractile form of tissue. It is one of the four major tissue types, the other three being epithelium, connective tissue and nervous tissue. Muscle contraction is used to move parts of the body, as well as to move substances within the body.
Cell division	Cell division is the process by which a cell (called the parent cell) divides into two cells (called daughter cells). Cell division is usually a small segment of a larger cell cycle. In meiosis, however, a cell is permanently transformed and cannot divide again.
Bladder	A hollow muscular storage organ for storing urine is a bladder.
Cancer	Cancer is a class of diseases or disorders characterized by uncontrolled division of cells and the ability of these cells to invade other tissues, either by direct growth into adjacent tissue through invasion or by implantation into distant sites by metastasis.
Inhibitor	An inhibitor is a type of effector (biology) that decreases or prevents the rate of a chemical reaction. They are often called negative catalysts.

Go to **Cram101.com** for the Practice Tests for this Chapter.

Growth factor	Growth factor is a protein that acts as a signaling molecule between cells (like cytokines and hormones) that attaches to specific receptors on the surface of a target cell and promotes differentiation and maturation of these cells.
Nucleus	In cell biology, the nucleus is found in all eukaryotic cells that contains most of the cell's genetic material. The nucleus has two primary functions: to control chemical reactions within the cytoplasm and to store information needed for cellular division.
Glycogen	Glycogen refers to a complex, extensively branched polysaccharide of many glucose monomers; serves as an energy-storage molecule in liver and muscle cells.
Ion	Ion refers to an atom or molecule that has gained or lost one or more electrons, thus acquiring an electrical charge.
Leaf	In botany, a leaf is an above-ground plant organ specialized for photosynthesis. For this purpose, a leaf is typically flat (laminar) and thin, to expose the chloroplast containing cells (chlorenchyma tissue) to light over a broad area, and to allow light to penetrate fully into the tissues.
Artery	Vessel that takes blood away from the heart to the tissues and organs of the body is called an artery.
Smooth muscle	Smooth muscle is a type of non-striated muscle, found within the "walls" of hollow organs; such as blood vessels, the bladder, the uterus, and the gastrointestinal tract. Smooth muscle is used to move matter within the body, via contraction; it generally operates "involuntarily", without nerve stimulation.
Second messenger	An intermediary compound that couples extracellular signals to intracellular processes and also amplifies a hormonal signal is referred to as second messenger.
Nerve	A nerve is an enclosed, cable-like bundle of nerve fibers or axons, which includes the glia that ensheath the axons in myelin.
Gene	Gene refers to a discrete unit of hereditary information consisting of a specific nucleotide sequence in DNA . Most of the genes of a eukaryote are located in its chromosomal DNA; a few are carried by the DNA of mitochondria and chloroplasts.
Nerve cell	A cell specialized to originate or transmit nerve impulses is referred to as nerve cell.
Transcription	Transcription is the process through which a DNA sequence is enzymatically copied by an RNA polymerase to produce a complementary RNA. Or, in other words, the transfer of genetic information from DNA into RNA.
Phytochrome	Phytochrome refers to photoreversible plant pigment that plants use to detect light and is involved in photoperiodism and other responses of plants such as etiolation.
Lens	The lens or crystalline lens is a transparent, biconvex structure in the eye that, along with the cornea, helps to refract light to focus on the retina. Its function is thus similar to a man-made optical lens.
Tissue	Group of similar cells which perform a common function is called tissue.
Plasmodesma	The connecting strands of protoplasm between the cytoplasm of adjacent cells which forms canals through the cell walls is plasmodesma. It may contain a desmotubule which links the endoplasmic reticulum of the adjacent cells.
Diffusion	Diffusion refers to the spontaneous movement of particles of any kind from where they are more concentrated to where they are less concentrated.
Mesophyll	Mesophyll refers to the green tissue in the interior of a leaf; a leaf's ground tissue system, the main site of photosynthesis.

Go to **Cram101.com** for the Practice Tests for this Chapter.
And, **NEVER** highlight a book again!

Phloem	In vascular plants, phloem is the living tissue that carries organic nutrients, particularly sucrose, to all parts of the plant where needed. In trees, the phloem is underneath and difficult to distinguish from bark,
Organ	Organ refers to a structure consisting of several tissues adapted as a group to perform specific functions.

Embryo	Embryo refers to a developing stage of a multicellular organism. In humans, the stage in the development of offspring from the first division of the zygote until body structures begin to appear, about the ninth week of gestation.
Cell division	Cell division is the process by which a cell (called the parent cell) divides into two cells (called daughter cells). Cell division is usually a small segment of a larger cell cycle. In meiosis, however, a cell is permanently transformed and cannot divide again.
Cell	The cell is the structural and functional unit of all living organisms, and is sometimes called the "building block of life."
Multicellular	Multicellular organisms are those organisms consisting of more than one cell, and having differentiated cells that perform specialized functions. Most life that can be seen with the naked eye is multicellular, as are all animals (i.e. members of the kingdom Animalia) and plants (i.e. members of the kingdom Plantae).
Morphogenesis	Morphogenesis is one of three fundamental aspects of developmental biology along with the control of cell growth and cellular differentiation. Morphogenesis is concerned with the shapes of tissues, organs and entire organisms and the positions of the various specialized cell types.
Pattern formation	During embryonic development, the emergence of a spatial organization in which the tissues and organs of the organism are all in their correct places is referred to as pattern formation.
Tissue	Group of similar cells which perform a common function is called tissue.
Protein	A protein is a complex, high-molecular-weight organic compound that consists of amino acids joined by peptide bonds. They are essential to the structure and function of all living cells and viruses. Many are enzymes or subunits of enzymes.
Keratin	Keratin is a family of fibrous structural proteins; tough and insoluble, they form the hard but nonmineralized structures found in reptiles, birds and mammals.
Genome	The genome of an organism is the whole hereditary information of an organism that is encoded in the DNA (or, for some viruses, RNA). This includes both the genes and the non-coding sequences. The genome of an organism is a complete DNA sequence of one set of chromosomes.
Nerve	A nerve is an enclosed, cable-like bundle of nerve fibers or axons, which includes the glia that ensheath the axons in myelin.
Clone	A group of genetically identical cells or organisms derived by asexual reproduction from a single parent is called a clone.
Species	Group of similarly constructed organisms capable of interbreeding and producing fertile offspring is a species.
In vitro fertilization	Uniting sperm and egg in a laboratory container, followed by the placement of a resulting early embryo in the mother's uterus is referred to as in vitro fertilization.
Fertilization	Fertilization is fusion of gametes to form a new organism. In animals, the process involves a sperm fusing with an ovum, which eventually leads to the development of an embryo.
In vitro	In vitro is an experimental technique where the experiment is performed in a test tube, or generally outside a living organism or cell.
Cloning	Cloning is the process of creating an identical copy of an original.
Cell cycle	An orderly sequence of events that extends from the time a eukaryotic cell divides to form two daughter cells to the time those daughter cells divide again is called cell cycle.
Tumor	An abnormal mass of cells that forms within otherwise normal tissue is a tumor. This growth

	can be either malignant or benign
Stem	Stem refers to that part of a plant's shoot system that supports the leaves and reproductive structures.
Stem cell	A stem cell is a primal undifferentiated cell which retains the ability to differentiate into other cell types. This ability allows them to act as a repair system for the body, replenishing other cells as long as the organism is alive.
Totipotent	Cell that has the full genetic potential of the organism and has the potential to develop into a complete organism is referred to as totipotent.
Embryonic stem cell	An embryonic stem cell is a cultured cell obtained from the undifferentiated inner mass cells of an early stage human embryo (sometimes called a blastocyst, which is an embryo that is between 50 to 150 cells).
Egg	An egg is the zygote, resulting from fertilization of the ovum. It nourishes and protects the embryo.
Muscle	Muscle is a contractile form of tissue. It is one of the four major tissue types, the other three being epithelium, connective tissue and nervous tissue. Muscle contraction is used to move parts of the body, as well as to move substances within the body.
Gene	Gene refers to a discrete unit of hereditary information consisting of a specific nucleotide sequence in DNA . Most of the genes of a eukaryote are located in its chromosomal DNA; a few are carried by the DNA of mitochondria and chloroplasts.
Promoter	In genetics, a promoter is a DNA sequence that enables a gene to be transcribed. The promoter is recognized by RNA polymerase, which then initiates transcription.
Segregation	The separation of homologous chromosomes during mitosis and meiosis. Known as Mendel's theory of Segregation.
Lens	The lens or crystalline lens is a transparent, biconvex structure in the eye that, along with the cornea, helps to refract light to focus on the retina. Its function is thus similar to a man-made optical lens.
Vesicle	In cell biology, a vesicle is a relatively small and enclosed compartment, separated from the cytosol by at least one lipid bilayer.
Gene expression	Gene expression is the process by which a gene's information is converted into the structures and functions of a cell. Gene expression is a multi-step process that begins with transcription, post transcriptional modification and translation, followed by folding, post-translational modification and targeting.
Growth factor	Growth factor is a protein that acts as a signaling molecule between cells (like cytokines and hormones) that attaches to specific receptors on the surface of a target cell and promotes differentiation and maturation of these cells.
Transcription	Transcription is the process through which a DNA sequence is enzymatically copied by an RNA polymerase to produce a complementary RNA. Or, in other words, the transfer of genetic information from DNA into RNA.
Cancer	Cancer is a class of diseases or disorders characterized by uncontrolled division of cells and the ability of these cells to invade other tissues, either by direct growth into adjacent tissue through invasion or by implantation into distant sites by metastasis.
Sepal	Outermost, sterile, leaflike covering of the flower is referred to as a sepal. A sepal is a tepal (a segment) of the calyx of a flower.
Acid	An acid is a water-soluble, sour-tasting chemical compound that when dissolved in water,

Go to **Cram101.com** for the Practice Tests for this Chapter.

gives a solution with a pH of less than 7.

Transcription factor	In molecular biology, a transcription factor is a protein that binds DNA at a specific promoter or enhancer region or site, where it regulates transcription.
Mutation	Mutation refers to a change in the nucleotide sequence of DNA; the ultimate source of genetic diversity.
Organ	Organ refers to a structure consisting of several tissues adapted as a group to perform specific functions.
Petal	A petal is one member or part of the corolla of a flower. It is the inner part of the perianth that comprises the sterile parts of a flower and consists of inner and outer tepals.
Sucrose	A disaccharide composed of glucose and fructose is called sucrose.
Larva	A free-living, sexually immature form in some animal life cycles that may differ from the adult in morphology, nutrition, and habitat is called larva.
Homozygous	When an organism is referred to as being homozygous for a specific gene, it means that it carries two identical copies of that gene for a given trait on the two corresponding chromosomes.
Cytoplasm	Cytoplasm refers to everything inside a cell between the plasma membrane and the nucleus; consists of a semifluid medium and organelles.
Segmentation	Segmentation in biology refers to the division of some metazoan bodies and plant body plans into a series of semi-repetitive segments, and the question of the benefits and costs of doing so.
Nucleus	In cell biology, the nucleus is found in all eukaryotic cells that contains most of the cell's genetic material. The nucleus has two primary functions: to control chemical reactions within the cytoplasm and to store information needed for cellular division.
Homeotic gene	A homeotic gene can, when modified cause homeotic mutations or bizarre and fundamental developemental reorganizations of the body. Homeotic genes in general transform parts of the body into structures appropriate to other positions.
Chromosome	A chromosome is, minimally, a very long, continuous piece of DNA, which contains many genes, regulatory elements and other intervening nucleotide sequences.
Prediction	Step of the scientific method which follows the formulation of a hypothesis and assists in creating the experimental design is referred to as prediction.
Nucleic acid	A nucleic acid is a complex, high-molecular-weight biochemical macromolecule composed of nucleotide chains that convey genetic information. The most common are deoxyribonucleic acid (DNA) and ribonucleic acid (RNA). They are found in all living cells and viruses.
Ventral	The surface or side of the body normally oriented upwards, away from the pull of gravity, is the dorsal side; the opposite side, typically the one closest to the ground when walking on all legs, swimming or flying, is the ventral side.
Vertebrate	Vertebrate is a subphylum of chordates, specifically, those with backbones or spinal columns. They started to evolve about 530 million years ago during the Cambrian explosion, which is part of the Cambrian period.
Iris	The colored part of the vertebrate eye, formed by the anterior portion of the choroid is called the iris.
Dorsal	In anatomy, the dorsal is the side in which the backbone is located. This is usually the top of an animal, although in humans it refers to the back.

Host	Host is an organism that harbors a parasite, mutual partner, or commensal partner; or a cell infected by a virus.
DNA replication	DNA replication is the process of copying a double-stranded DNA strand in a cell, prior to cell division. The two resulting double strands are identical (if the replication went well), and each of them consists of one original and one newly synthesized strand.
Bacteria	The domain that contains procaryotic cells with primarily diacyl glycerol diesters in their membranes and with bacterial rRNA. Bacteria also is a general term for organisms that are composed of procaryotic cells and are not multicellular.
Nucleotide	A nucleotide is a chemical compound that consists of a heterocyclic base, a sugar, and one or more phosphate groups. In the most common nucleotides the base is a derivative of purine or pyrimidine, and the sugar is pentose - deoxyribose or ribose. They are the structural units of RNA and DNA.
Enzyme	An enzyme is a protein that catalyzes, or speeds up, a chemical reaction. They are essential to sustain life because most chemical reactions in biological cells would occur too slowly, or would lead to different products, without them.
Radioactive	A term used to describe the property of releasing energy or particles from an unstable atom is called radioactive.
Hybridization	In molecular biology hybridization is the process of joining two complementary strands of
Restriction enzyme	Restriction enzyme refers to a bacterial enzyme that cuts up foreign DNA, thus protecting bacteria against intruding DNA from phages and other organisms. They are used in DNA technology to cut DNA molecules in reproducible ways.
Recombinant DNA	Recombinant DNA is an artificial DNA sequence resulting from the combining of two other DNA sequences in a plasmid.
Molecular biology	Molecular biology overlaps with other areas of biology and chemistry, particularly genetics and biochemistry. Molecular biology chiefly concerns itself with understanding the interactions between the various systems of a cell, including the interrelationship of DNA, RNA and protein synthesis and learning how these interactions are regulated.
Biology	Biology is the branch of science dealing with the study of life. It is concerned with the characteristics, classification, and behaviors of organisms, how species come into existence, and the interactions they have with each other and with the environment.
Gene	Gene refers to a discrete unit of hereditary information consisting of a specific nucleotide sequence in DNA . Most of the genes of a eukaryote are located in its chromosomal DNA; a few are carried by the DNA of mitochondria and chloroplasts.
Cell division	Cell division is the process by which a cell (called the parent cell) divides into two cells (called daughter cells). Cell division is usually a small segment of a larger cell cycle. In meiosis, however, a cell is permanently transformed and cannot divide again.
Cell	The cell is the structural and functional unit of all living organisms, and is sometimes called the "building block of life."
Genome	The genome of an organism is the whole hereditary information of an organism that is encoded in the DNA (or, for some viruses, RNA). This includes both the genes and the non-coding sequences. The genome of an organism is a complete DNA sequence of one set of chromosomes.
Eukaryote	A eukaryote is an organism with complex cells, in which the genetic material is organized into membrane-bound nuclei. They comprise animals, plants, and fungi—which are mostly multicellular—as well as various other groups that are collectively classified as protists.
Transgenic	Referring to an animal or a plant that expresses DNA derived from another species is called

Go to **Cram101.com** for the Practice Tests for this Chapter.

transgenic.

Vector	A vector is an organism that does not cause disease itself but which spreads infection by conveying pathogens from one host to another.
Plasmid	Plasmid is a circular double-stranded DNA molecule that is separate from the chromosomal DNA. They usually occur in bacteria and often contain genes that confer a selective advantage to the bacterium harboring them, e.g., the ability to make the bacterium antibiotic resistant.
Salt	Salt is a term used for ionic compounds composed of positively charged cations and negatively charged anions, so that the product is neutral and without a net charge.
Cell wall	Cell wall refers to a protective layer external to the plasma membrane in plant cells, bacteria, fungi, and some protists; protects the cell and helps maintain its shape.
Lipid	Lipid is one class of aliphatic hydrocarbon-containing organic compounds essential for the structure and function of living cells. They are characterized by being water-insoluble but soluble in nonpolar organic solvents.
Population	Group of organisms of the same species occupying a certain area and sharing a common gene pool is referred to as population.
Antibiotic resistance	The ability of a mutated pathogen to resist the effects of an antibiotic that normally kills it is an antibiotic resistance.
Antibiotic	Antibiotic refers to substance such as penicillin or streptomycin that is toxic to microorganisms. Usually a product of a particular microorvanism or plant.
Protein	A protein is a complex, high-molecular-weight organic compound that consists of amino acids joined by peptide bonds. They are essential to the structure and function of all living cells and viruses. Many are enzymes or subunits of enzymes.
Chromosome	A chromosome is, minimally, a very long, continuous piece of DNA, which contains many genes, regulatory elements and other intervening nucleotide sequences.
Tissue	Group of similar cells which perform a common function is called tissue.
Codon	Codon refers to a three-nucleotide sequence in mRNA that specifies a particular amino acid or polypeptide termination signal; the basic unit of the genetic code.
Tertiary structure	Tertiary structure refers to the complex three-dimensional structure of a single peptide chain; held in place by disulfide bonds between cysteines.
Polypeptide	Polypeptide refers to polymer of many amino acids linked by peptide bonds.
Recombination	Genetic recombination is the transmission-genetic process by which the combinations of alleles observed at different loci in two parental individuals become shuffled in offspring individuals.
Genetic marker	A genetic marker is a specific piece of DNA with a known position on the genome. It is a genetic technique to follow a certain disease or gene. It is known that pieces of DNA that lie near each other, tend to be inherited to the next organism together.
Stem cell	A stem cell is a primal undifferentiated cell which retains the ability to differentiate into other cell types. This ability allows them to act as a repair system for the body, replenishing other cells as long as the organism is alive.
Stem	Stem refers to that part of a plant's shoot system that supports the leaves and reproductive structures.
Phenotype	The phenotype of an individual organism is either its total physical appearance and constitution or a specific manifestation of a trait, such as size or eye color, that varies

between individuals. It is determined to some extent by genotype.

Gene expression	Gene expression is the process by which a gene's information is converted into the structures and functions of a cell. Gene expression is a multi-step process that begins with transcription, post transcriptional modification and translation, followed by folding, post-translational modification and targeting.
Skin	Skin is an organ of the integumentary system composed of a layer of tissues that protect underlying muscles and organs.
Reverse transcriptase	Reverse transcriptase is an enzyme that is able to transcribe RNA into DNA. That is, reverse transcriptase is able to copy genetic information from RNA to DNA, which is the reverse of the more typical direction
Transcriptase	Transcriptase refers to an enzyme that catalyzes transcription; in viruses with RNA genomes. This enzyme is an RNA-dependent RNA polymerase that is used to make RNA copies of the RNA genomes.
Glycerol	Glycerol is a three-carbon substance that forms the backbone of fatty acids in fats. When the body uses stored fat as a source of energy, glycerol and fatty acids are released into the bloodstream. The glycerol component can be converted to glucose by the liver and provides energy for cellular metabolism.
Transcription factor	In molecular biology, a transcription factor is a protein that binds DNA at a specific promoter or enhancer region or site, where it regulates transcription.
Transcription	Transcription is the process through which a DNA sequence is enzymatically copied by an RNA polymerase to produce a complementary RNA. Or, in other words, the transfer of genetic information from DNA into RNA.
Promoter	In genetics, a promoter is a DNA sequence that enables a gene to be transcribed. The promoter is recognized by RNA polymerase, which then initiates transcription.
Blood	Blood is a circulating tissue composed of fluid plasma and cells. The main function of blood is to supply nutrients (oxygen, glucose) and constitutional elements to tissues and to remove waste products.
Kidney	The kidney is a bean-shaped excretory organ in vertebrates. Part of the urinary system, the kidneys filter wastes (especially urea) from the blood and excrete them, along with water, as urine.
Dialysis	Dialysis refers to separation and disposal of metabolic wastes from the blood by mechanical means; an artificial method of performing the functions of the kidneys.
Diabetes mellitus	Diabetes mellitus is a medical disorder characterized by varying or persistent hyperglycemia (elevated blood sugar levels), especially after eating. All types of diabetes mellitus share similar symptoms and complications at advanced stages.
Hormone	A hormone is a chemical messenger from one cell to another. All multicellular organisms produce hormones. The best known hormones are those produced by endocrine glands of vertebrate animals, but hormones are produced by nearly every organ system and tissue type in a human or animal body. Hormone molecules are secreted directly into the bloodstream, they move by circulation or diffusion to their target cells, which may be nearby cells in the same tissue or cells of a distant organ of the body.
Insulin	Insulin is a polypeptide hormone that regulates carbohydrate metabolism. Apart from being the primary effector in carbohydrate homeostasis, it also has a substantial effect on small vessel muscle tone, controls storage and release of fat (triglycerides) and cellular uptake of both amino acids and some electrolytes.

Go to **Cram101.com** for the Practice Tests for this Chapter.

Crop	An organ, found in both earthworms and birds, in which ingested food is temporarily stored before being passed to the gizzard, where it is pulverized is the crop.
Genetics	Genetics is the science of genes, heredity, and the variation of organisms.
Cloning	Cloning is the process of creating an identical copy of an original.
Ecosystem	In general terms an ecosystem can be thought of as an assemblage of organisms (plant, animal and other living organisms living together with their environment, functioning as a loose unit. That is, a dynamic and complex whole, interacting as an "ecological unit".
Larva	A free-living, sexually immature form in some animal life cycles that may differ from the adult in morphology, nutrition, and habitat is called larva.
Biotechnology	Biotechnology refers to the use of living organisms to perform useful tasks; today, usually involves DNA technology.
Cystic fibrosis	Cystic fibrosis refers to a genetic disease that occurs in people with two copies of a certain recessive allele; characterized by an excessive secretion of mucus and consequent vulnerability to infection; fatal if untreated.
Inhibitor	An inhibitor is a type of effector (biology) that decreases or prevents the rate of a chemical reaction. They are often called negative catalysts.
Vitamin	A Vitamin is an organic molecule required by a living organism in minute amounts for proper health. An organism deprived of all sources of a particular vitamin will eventually suffer from disease symptoms specific to that vitamin.
Species	Group of similarly constructed organisms capable of interbreeding and producing fertile offspring is a species.
Human genome	The human genome is the genome of Homo sapiens. It is made up of 23 chromosome pairs with a total of about 3 billion DNA base pairs.
Repetitive DNA	Repetitive DNA refers to nucleotide sequences that are present in many copies in the DNA of a genome. The repeated sequences may be long or short and may be located next to each other or dispersed in the DNA.
Polymerase chain reaction	Polymerase chain reaction is a molecular biology technique for enzymatically replicating DNA without using a living organism, such as E. coli or yeast. The technique allows a small amount of the DNA molecule to be amplified many times, in an exponential manner.
Polymerase	DNA or RNA enzymes that catalyze the synthesis of nucleic acids on preexisting nucleic acid templates, assembling RNA from ribonucleotides or DNA from deoxyribonucleotides is referred to as polymerase.
Y chromosome	Male sex chromosome that carries genes involved in sex determination is referred to as the Y chromosome. It contains the genes that cause testis development, thus determining maleness.
Artificial selection	Artificial selection is the process of intentional or unintentional modification of a species through human actions which encourage the breeding of certain traits over others. When the process leads to undesirable outcome, it is called negative selection.
Pathogen	A pathogen or infectious agent is a biological agent that causes disease or illness to its host.The term is most often used for agents that disrupt the normal physiology of a multicellular animal or plant.
Sickle-cell anemia	Sickle-cell anemia refers to a disease caused by a point mutation. This malfunction produces sickle-shaped red blood cells. Sickle-cell anemia is the name of a specific form of sickle cell disease in which there is homozygosity for the mutation that causes Hgb S.
Anemia	Anemia is a deficiency of red blood cells and/or hemoglobin. This results in a reduced

ability of blood to transfer oxygen to the tissues, and this causes hypoxia; since all human cells depend on oxygen for survival, varying degrees of anemia can have a wide range of clinical consequences.

Go to **Cram101.com** for the Practice Tests for this Chapter.

Hemoglobin	Hemoglobin is the iron-containing oxygen-transport metalloprotein in the red cells of the blood in mammals and other animals. Hemoglobin transports oxygen from the lungs to the rest of the body, such as to the muscles, where it releases the oxygen load.
Gene	Gene refers to a discrete unit of hereditary information consisting of a specific nucleotide sequence in DNA . Most of the genes of a eukaryote are located in its chromosomal DNA; a few are carried by the DNA of mitochondria and chloroplasts.
Pigment	Pigment is any material resulting in color in plant or animal cells which is the result of selective absorption.
Protein	A protein is a complex, high-molecular-weight organic compound that consists of amino acids joined by peptide bonds. They are essential to the structure and function of all living cells and viruses. Many are enzymes or subunits of enzymes.
Amino acid	An amino acid is any molecule that contains both amino and carboxylic acid functional groups. They are the basic structural building units of proteins. They form short polymer chains called peptides or polypeptides which in turn form structures called proteins.
Acid	An acid is a water-soluble, sour-tasting chemical compound that when dissolved in water, gives a solution with a pH of less than 7.
Blood	Blood is a circulating tissue composed of fluid plasma and cells. The main function of blood is to supply nutrients (oxygen, glucose) and constitutional elements to tissues and to remove waste products.
Low-density lipoprotein	Low-density lipoprotein refers to a class and range of lipoprotein particles, varying in their size (18-25 nm in diameter) and contents, which carry cholesterol in the blood and around the body, for use by cells. It is the final stage of VLDL (very low-density lipoprotein) which is produced by the liver.
Lipoprotein	A lipoprotein is a biochemical assembly that contains both proteins and lipids and may be structural or catalytic in function. They may be enzymes, proton pumps, ion pumps, or some combination of these functions.
Receptor	A receptor is a protein on the cell membrane or within the cytoplasm or cell nucleus that binds to a specific molecule (a ligand), such as a neurotransmitter, hormone, or other substance, and initiates the cellular response to the ligand. Receptor, in immunology, the region of an antibody which shows recognition of an antigen.
Homozygous	When an organism is referred to as being homozygous for a specific gene, it means that it carries two identical copies of that gene for a given trait on the two corresponding chromosomes.
Allele	An allele is any one of a number of viable DNA codings of the same gene (sometimes the term refers to a non-gene sequence) occupying a given locus (position) on a chromosome.
Anemia	Anemia is a deficiency of red blood cells and/or hemoglobin. This results in a reduced ability of blood to transfer oxygen to the tissues, and this causes hypoxia; since all human cells depend on oxygen for survival, varying degrees of anemia can have a wide range of clinical consequences.
Hypercholest-rolemia	Hypercholesterolemia is the presence of high levels of cholesterol in the blood. It is not a disease but a metabolic derangement that can be secondary to many diseases and can contribute to many forms of disease, most notably cardiovascular disease.
Receptor protein	Protein located in the plasma membrane or within the cell that binds to a substance that alters some metabolic aspect of the cell is referred to as receptor protein. It will only link up with a substance that has a certain shape that allows it to bind to the receptor.

Cystic fibrosis	Cystic fibrosis refers to a genetic disease that occurs in people with two copies of a certain recessive allele; characterized by an excessive secretion of mucus and consequent vulnerability to infection; fatal if untreated.
Liver	The liver is an organ in vertebrates, including humans. It plays a major role in metabolism and has a number of functions in the body including drug detoxification, glycogen storage, and plasma protein synthesis. It also produces bile, which is important for digestion.
Ion	Ion refers to an atom or molecule that has gained or lost one or more electrons, thus acquiring an electrical charge.
Respiratory system	The respiratory system is the biological system of any organism that engages in gas exchange.In humans and other mammals, the respiratory system consists of the airways, the lungs, and the respiratory muscles that mediate the movement of air into and out of the body.
Hemophilia	Hemophilia is the name of any of several hereditary genetic illnesses that impair the body's ability to control bleeding. Genetic deficiencies cause lowered plasma clotting factor activity so as to compromise blood-clotting; when a blood vessel is injured, a scab will not form and the vessel can continue to bleed excessively for a very long period of time.
Prion	Prion is an infectious particle consisting of protein only and no nucleic acid which is believed to be linked to several diseases of the central nervous system.
Kuru	Kuru, a disease, related to bovine spongiform encephalopathy (BSE), once affected a small subgroup of the Fore people of highland New Guinea as a result of ritual cannibalism.
Brain	The part of the central nervous system involved in regulating and controlling body activity and interpreting information from the senses transmitted through the nervous system is referred to as the brain.
Sickle-cell anemia	Sickle-cell anemia refers to a disease caused by a point mutation. This malfunction produces sickle-shaped red blood cells. Sickle-cell anemia is the name of a specific form of sickle cell disease in which there is homozygosity for the mutation that causes Hgb S.
Heterozygous	Heterozygous means that the organism carries a different version of that gene on each of the two corresponding chromosomes.
Biochemical pathway	A chain of chemical reactions that occur in living things to produce a chemical compound is referred to as biochemical pathway.
X chromosome	The X chromosome is the female sex chromosome that carries genes involved in sex determination. Females have two X chromosomes, while males have one X and one Y chromosome.
Chromosome	A chromosome is, minimally, a very long, continuous piece of DNA, which contains many genes, regulatory elements and other intervening nucleotide sequences.
Phenotype	The phenotype of an individual organism is either its total physical appearance and constitution or a specific manifestation of a trait, such as size or eye color, that varies between individuals. It is determined to some extent by genotype.
Deletion	Deletion refers to the loss of one or more nucleotides from a gene by mutation; the loss of a fragment of a chromosome.
Cloning	Cloning is the process of creating an identical copy of an original.
Pedigree	A record of one's ancestors, offspring, siblings, and their offspring that may be used to determine the pattern of certain genes or disease inheritance within a family is a pedigree.
Human genome	The human genome is the genome of Homo sapiens. It is made up of 23 chromosome pairs with a total of about 3 billion DNA base pairs.
Genome	The genome of an organism is the whole hereditary information of an organism that is encoded

in the DNA (or, for some viruses, RNA). This includes both the genes and the non-coding sequences. The genome of an organism is a complete DNA sequence of one set of chromosomes.

Enzyme	An enzyme is a protein that catalyzes, or speeds up, a chemical reaction. They are essential to sustain life because most chemical reactions in biological cells would occur too slowly, or would lead to different products, without them.
Mutation	Mutation refers to a change in the nucleotide sequence of DNA; the ultimate source of genetic diversity.
Biology	Biology is the branch of science dealing with the study of life. It is concerned with the characteristics, classification, and behaviors of organisms, how species come into existence, and the interactions they have with each other and with the environment.
Translation	Translation is the second process of protein biosynthesis. In translation, messenger RNA is decoded to produce a specific polypeptide according to the rules specified by the genetic code.
Cytosine	Cytosine is one of the 5 main nucleobases used in storing and transporting genetic information within a cell in the nucleic acids DNA and RNA. It is a pyrimidine derivative, with a heterocyclic aromatic ring and two substituents attached. The nucleoside of cytosine is cytidine.
Fertilization	Fertilization is fusion of gametes to form a new organism. In animals, the process involves a sperm fusing with an ovum, which eventually leads to the development of an embryo.
Imprinting	Learning that is limited to a specific critical period in an animal's life and that is generally irreversible is called imprinting.
Cell	The cell is the structural and functional unit of all living organisms, and is sometimes called the "building block of life."
Embryo	Embryo refers to a developing stage of a multicellular organism. In humans, the stage in the development of offspring from the first division of the zygote until body structures begin to appear, about the ninth week of gestation.
Bacteria	The domain that contains procaryotic cells with primarily diacyl glycerol diesters in their membranes and with bacterial rRNA. Bacteria also is a general term for organisms that are composed of procaryotic cells and are not multicellular.
Population	Group of organisms of the same species occupying a certain area and sharing a common gene pool is referred to as population.
Polymerase chain reaction	Polymerase chain reaction is a molecular biology technique for enzymatically replicating DNA without using a living organism, such as E. coli or yeast. The technique allows a small amount of the DNA molecule to be amplified many times, in an exponential manner.
Polymerase	DNA or RNA enzymes that catalyze the synthesis of nucleic acids on preexisting nucleic acid templates, assembling RNA from ribonucleotides or DNA from deoxyribonucleotides is referred to as polymerase.
Implantation	Implantation refers to attachment and penetration of the embryo into the lining of the uterus.
Restriction enzyme	Restriction enzyme refers to a bacterial enzyme that cuts up foreign DNA, thus protecting bacteria against intruding DNA from phages and other organisms. They are used in DNA technology to cut DNA molecules in reproducible ways.
Double helix	Double helix refers to the form of native DNA, referring to its two adjacent polynucleotide strands wound into a spiral shape.

Helix	A helix is a twisted shape like a spring, screw or a spiral staircase. They are important in biology, as DNA and many proteins have spiral substructures, known a alpha helix.
Cancer	Cancer is a class of diseases or disorders characterized by uncontrolled division of cells and the ability of these cells to invade other tissues, either by direct growth into adjacent tissue through invasion or by implantation into distant sites by metastasis.
Benign tumor	A benign tumor does not invade neighboring tissues and do not seed metastases, but may locally grow to great size. They usually do not return after surgical removal.
Tumor	An abnormal mass of cells that forms within otherwise normal tissue is a tumor. This growth can be either malignant or benign
Organ	Organ refers to a structure consisting of several tissues adapted as a group to perform specific functions.
Tissue	Group of similar cells which perform a common function is called tissue.
Malignant	In medicine, malignant is a clinical term that is used to describe a clinical course that progresses rapidly to death. It is typically applied to neoplasms that show aggressive behavior characterized by local invasion or distant metastasis.
Cancer cell	A cell that divides and reproduces abnormally and has the potential to spread throughout the body, crowding out normal cells and tissue is referred to as a cancer cell.
Skin	Skin is an organ of the integumentary system composed of a layer of tissues that protect underlying muscles and organs.
Virus	Obligate intracellular parasite of living cells consisting of an outer capsid and an inner core of nucleic acid is referred to as virus. The term virus usually refers to those particles that infect eukaryotes whilst the term bacteriophage or phage is used to describe those infecting prokaryotes.
Carcinogen	Carcinogen refers to a cancer-causing agent, either high-energy radiation or a chemical.
Proto-oncogene	Normal gene that can become an oncogene through mutation or increased expression is referred to as proto-oncogene.
Cell division	Cell division is the process by which a cell (called the parent cell) divides into two cells (called daughter cells). Cell division is usually a small segment of a larger cell cycle. In meiosis, however, a cell is permanently transformed and cannot divide again.
Colon	The colon is the part of the intestine from the cecum to the rectum. Its primary purpose is to extract water from feces.
Oncogene	An oncogene is a modified gene that increases the malignancy of a tumor cell. Some oncogenes, usually involved in early stages of cancer development, increase the chance that a normal cell develops into a tumor cell, possibly resulting in cancer.
Substrate	A substrate is a molecule which is acted upon by an enzyme. Each enzyme recognizes only the specific substrate of the reaction it catalyzes. A surface in or on which an organism lives.
Gene therapy	Gene therapy is the insertion of genes into an individual's cells and tissues to treat a disease, and hereditary diseases in particular.
Stem	Stem refers to that part of a plant's shoot system that supports the leaves and reproductive structures.
Metabolism	Metabolism is the biochemical modification of chemical compounds in living organisms and cells. This includes the biosynthesis of complex organic molecules (anabolism) and their breakdown (catabolism).

Phagocytosis	Phagocytosis is a form of endocytosis where large particles are enveloped by the cell membrane of a (usually larger) cell and internalized to form a phagosome, or "food vacuole."
Cell	The cell is the structural and functional unit of all living organisms, and is sometimes called the "building block of life."
Blood	Blood is a circulating tissue composed of fluid plasma and cells. The main function of blood is to supply nutrients (oxygen, glucose) and constitutional elements to tissues and to remove waste products.
Vertebrate	Vertebrate is a subphylum of chordates, specifically, those with backbones or spinal columns. They started to evolve about 530 million years ago during the Cambrian explosion, which is part of the Cambrian period.
Lymph	Lymph originates as blood plasma lost from the circulatory system, which leaks out into the surrounding tissues. The lymphatic system collects this fluid by diffusion into lymph capillaries, and returns it to the circulatory system.
Plasma	In physics and chemistry, a plasma is an ionized gas, and is usually considered to be a distinct phase of matter. "Ionized" in this case means that at least one electron has been dissociated from a significant fraction of the molecules.
Brain	The part of the central nervous system involved in regulating and controlling body activity and interpreting information from the senses transmitted through the nervous system is referred to as the brain.
Bone marrow	Bone marrow is the tissue comprising the center of large bones. It is the place where new blood cells are produced. Bone marrow contains two types of stem cells: hemopoietic (which can produce blood cells) and stromal (which can produce fat, cartilage and bone).
Tumor	An abnormal mass of cells that forms within otherwise normal tissue is a tumor. This growth can be either malignant or benign
Innate	Innate is used to describe an inherent or intrinsic characteristic or property of some thing, such as a quality or capability which is possessed since birth.
Bacteria	The domain that contains procaryotic cells with primarily diacyl glycerol diesters in their membranes and with bacterial rRNA. Bacteria also is a general term for organisms that are composed of procaryotic cells and are not multicellular.
Skin	Skin is an organ of the integumentary system composed of a layer of tissues that protect underlying muscles and organs.
Enzyme	An enzyme is a protein that catalyzes, or speeds up, a chemical reaction. They are essential to sustain life because most chemical reactions in biological cells would occur too slowly, or would lead to different products, without them.
Microorganism	A microorganism is an organism that is so small that it is microscopic (invisible to the naked eye). They are often illustrated using single-celled, or unicellular organisms; however, some unicellular protists are visible to the naked eye, and some multicellular species are microscopic.
Cilia	Numerous short, hairlike structures projecting from the cell surface that enable locomotion are cilia.
Stomach	The stomach is an organ in the alimentary canal used to digest food. It's primary function is not the absorption of nutrients from digested food; rather, the main job of the stomach is to break down large food molecules into smaller ones, so that they can be absorbed into the blood more easily.
Small intestine	The small intestine is the part of the gastrointestinal tract between the stomach and the

large intestine (colon). In humans over 5 years old it is about 7m long. It is divided into three structural parts: duodenum, jejunum and ileum.

Intestine	The intestine is the portion of the alimentary canal extending from the stomach to the anus and, in humans and mammals, consists of two segments, the small intestine and the large intestine. The intestine is the part of the body responsible for extracting nutrition from food.
Large intestine	In anatomy of the digestive system, the colon, also called the large intestine or large bowel, is the part of the intestine from the cecum ('caecum' in British English) to the rectum. Its primary purpose is to extract water from feces.
Complement	Complement is a group of proteins of the complement system, found in blood serum which act in concert with antibodies to achieve the destruction of non-self particles such as foreign blood cells or bacteria.
Species	Group of similarly constructed organisms capable of interbreeding and producing fertile offspring is a species.
Circulatory system	The circulatory system or cardiovascular system is the organ system which circulates blood around the body of most animals.
Lymph node	A lymph node acts as a filter, with an internal honeycomb of connective tissue filled with lymphocytes that collect and destroy bacteria and viruses. When the body is fighting an infection, these lymphocytes multiply rapidly and produce a characteristic swelling of the lymph node.
Node	In botany, a node is the place on a stem where a leaf is attached.
Histamine	Histamine is a biogenic amine chemical involved in local immune responses as well as regulating physiological function in the gut and acting as a neurotransmitter. Histamines also play a role in chemotaxis of white blood cells.
Tissue	Group of similar cells which perform a common function is called tissue.
Immune response	The body's defensive reaction to invasion by bacteria, viral agents, or other foreign substances is called the immune response.
Immune system	The immune system is the system of specialized cells and organs that protect an organism from outside biological influences. When the immune system is functioning properly, it protects the body against bacteria and viral infections, destroying cancer cells and foreign substances.
Antigen	An antigen is a substance that stimulates an immune response, especially the production of antibodies. They are usually proteins or polysaccharides, but can be any type of molecule, including small molecules (haptens) coupled to a protein (carrier).
Protein	A protein is a complex, high-molecular-weight organic compound that consists of amino acids joined by peptide bonds. They are essential to the structure and function of all living cells and viruses. Many are enzymes or subunits of enzymes.
Host	Host is an organism that harbors a parasite, mutual partner, or commensal partner; or a cell infected by a virus.
T cell	T cell refers to a type of lymphocyte that matures in the thymus and is responsible for cell-mediated immunity Every effective immune response involves T cell activation.
Pathogen	A pathogen or infectious agent is a biological agent that causes disease or illness to its host. The term is most often used for agents that disrupt the normal physiology of a multicellular animal or plant.

Go to **Cram101.com** for the Practice Tests for this Chapter.

Clonal selection	The production of a lineage of genetically identical cells that recognize and attack the specific antigen that stimulated their proliferation. Clonal selection is the mechanism that underlies the immune system's specificity and memory of antigens.
Effector	An effector is a molecule (originally referring to small molecules but now encompassing any regulatory molecule, includes proteins) that binds to a protein and thereby alters the activity of that protein.
Deletion	Deletion refers to the loss of one or more nucleotides from a gene by mutation; the loss of a fragment of a chromosome.
Polypeptide	Polypeptide refers to polymer of many amino acids linked by peptide bonds.
Amino acid	An amino acid is any molecule that contains both amino and carboxylic acid functional groups. They are the basic structural building units of proteins. They form short polymer chains called peptides or polypeptides which in turn form structures called proteins.
Acid	An acid is a water-soluble, sour-tasting chemical compound that when dissolved in water, gives a solution with a pH of less than 7.
Antibody	An antibody is a protein used by the immune system to identify and neutralize foreign objects like bacteria and viruses. Each antibody recognizes a specific antigen unique to its target.
Immunoglobulin	Immunoglobulin refers to a globular plasma protein that functions as an antibody.
Clone	A group of genetically identical cells or organisms derived by asexual reproduction from a single parent is called a clone.
Steroid	A steroid is a lipid characterized by a carbon skeleton with four fused rings. Different steroids vary in the functional groups attached to these rings. Hundreds of distinct steroids have been identified in plants and animals. Their most important role in most living systems is as hormones.
Estrogen	Estrogen is a steroid that functions as the primary female sex hormone. While present in both men and women, they are found in women in significantly higher quantities.
Monoclonal antibody	Monoclonal antibody refers to an antibody secreted by a clone of cells and consequently specific for the one antigen that triggered the development of the clone.
Cancer	Cancer is a class of diseases or disorders characterized by uncontrolled division of cells and the ability of these cells to invade other tissues, either by direct growth into adjacent tissue through invasion or by implantation into distant sites by metastasis.
Cancer cell	A cell that divides and reproduces abnormally and has the potential to spread throughout the body, crowding out normal cells and tissue is referred to as a cancer cell.
Immunization	Use of a vaccine to protect the body against specific disease-causing agents is called immunization.
Gene	Gene refers to a discrete unit of hereditary information consisting of a specific nucleotide sequence in DNA . Most of the genes of a eukaryote are located in its chromosomal DNA; a few are carried by the DNA of mitochondria and chloroplasts.
Leukocyte	A white blood cell is a leukocyte. They help to defend the body against infectious disease and foreign materials as part of the immune system.
Organ	Organ refers to a structure consisting of several tissues adapted as a group to perform specific functions.
Mammal	Homeothermic vertebrate characterized especially by the presence of hair and mammary glands is a mammal.

Go to **Cram101.com** for the Practice Tests for this Chapter.

Allele	An allele is any one of a number of viable DNA codings of the same gene (sometimes the term refers to a non-gene sequence) occupying a given locus (position) on a chromosome.
Variable region	Variable region refers to the part of an antibody molecule that differs among antibodies; the ends of the variable regions of the light and heavy chains form the specific binding site for antigens.
Nucleic acid	A nucleic acid is a complex, high-molecular-weight biochemical macromolecule composed of nucleotide chains that convey genetic information. The most common are deoxyribonucleic acid (DNA) and ribonucleic acid (RNA). They are found in all living cells and viruses.
Transcription	Transcription is the process through which a DNA sequence is enzymatically copied by an RNA polymerase to produce a complementary RNA. Or, in other words, the transfer of genetic information from DNA into RNA.
Constant region	Constant region refers to the part of an antibody with similar amino acid sequences in all antibodies of the same type.
Invertebrate	Invertebrate is a term coined by Jean-Baptiste Lamarck to describe any animal without a spinal column. It therefore includes all animals except vertebrates (fish, reptiles, amphibians, birds and mammals).
Complement system	Group of plasma proteins that form a nonspecific defense mechanism, often by puncturing microbes, is called the complement system. The complement system is derived from many small plasma proteins that form the complex biochemical cascade of the immune system.
Molecule	A molecule is the smallest particle of a pure chemical substance that still retains its chemical composition and properties.
Arthritis	Arthritis is a group of conditions that affect the health of the bone joints in the body. One in three adult Americans suffer from some form of arthritis and the disease affects about twice as many women as men.
Nervous system	The nervous system of an animal coordinates the activity of the muscles, monitors the organs, constructs and processes input from the senses, and initiates actions.
Virus	Obligate intracellular parasite of living cells consisting of an outer capsid and an inner core of nucleic acid is referred to as virus. The term virus usually refers to those particles that infect eukaryotes whilst the term bacteriophage or phage is used to describe those infecting prokaryotes.
Retrovirus	A retrovirus is a virus which has a genome consisting of two RNA molecules, which may or may not be identical. It relies on the enzyme reverse transcriptase to perform the reverse transcription of its genome from RNA into DNA, which can then be integrated into the host's genome with an integrase enzyme.
Plasma membrane	Membrane surrounding the cytoplasm that consists of a phospholipid bilayer with embedded proteins is referred to as plasma membrane.
Receptor	A receptor is a protein on the cell membrane or within the cytoplasm or cell nucleus that binds to a specific molecule (a ligand), such as a neurotransmitter, hormone, or other substance, and initiates the cellular response to the ligand. Receptor, in immunology, the region of an antibody which shows recognition of an antigen.
Genome	The genome of an organism is the whole hereditary information of an organism that is encoded in the DNA (or, for some viruses, RNA). This includes both the genes and the non-coding sequences. The genome of an organism is a complete DNA sequence of one set of chromosomes.
Protease	Protease refers to an enzyme that breaks peptide bonds between amino acids of proteins.
Molecular	Molecular biology overlaps with other areas of biology and chemistry, particularly genetics

Go to **Cram101.com** for the Practice Tests for this Chapter.

biology	and biochemistry. Molecular biology chiefly concerns itself with understanding the interactions between the various systems of a cell, including the interrelationship of DNA, RNA and protein synthesis and learning how these interactions are regulated.
Biology	Biology is the branch of science dealing with the study of life. It is concerned with the characteristics, classification, and behaviors of organisms, how species come into existence, and the interactions they have with each other and with the environment.

Evolution	In biology, evolution is the process by which novel traits arise in populations and are passed on from generation to generation. Its action over large stretches of time explains the origin of new species and ultimately the vast diversity of the biological world.
Fossil	A preserved remnant or impression of an organism that lived in the past is referred to as fossil.
Isotope	An isotope is a form of an element whose nuclei have the same atomic number - the number of protons in the nucleus - but different mass numbers because they contain different numbers of neutrons.
Radioactive	A term used to describe the property of releasing energy or particles from an unstable atom is called radioactive.
Half-life	The half-life of a quantity subject to exponential decay is the time required for the quantity to fall to half of its initial value.
Unicellular	Unicellular organisms carry out all the functions of life. Unicellular species are those whose members consist of a single cell throughout their life cycle. This latter qualification is significant since most multicellular organisms consist of a single cell at the beginning of their life cycles.
Eukaryotic cell	Eukaryotic cell refers to a type of cell that has a membrane-enclosed nucleus and other membrane enclosed organelles. All organisms except bacteria and archaea are composed of eukaryotic cells.
Cell	The cell is the structural and functional unit of all living organisms, and is sometimes called the "building block of life."
Multicellular	Multicellular organisms are those organisms consisting of more than one cell, and having differentiated cells that perform specialized functions. Most life that can be seen with the naked eye is multicellular, as are all animals (i.e. members of the kingdom Animalia) and plants (i.e. members of the kingdom Plantae).
Mantle	The mantle is an organ found in mollusks. It is the dorsal body wall covering the main body, or visceral mass. The epidermis of this organ secretes calcium carbonate to create a shell.
Climate	Weather condition of an area including especially prevailing temperature and average daily/yearly rainfall over a long period of time is called climate.
Species	Group of similarly constructed organisms capable of interbreeding and producing fertile offspring is a species.
Photosynthesis	Photosynthesis is a biochemical process in which plants, algae, and some bacteria harness the energy of light to produce food. Ultimately, nearly all living things depend on energy produced from photosynthesis for their nourishment, making it vital to life on Earth.
Plankton	Plankton are drifting organisms that inhabit the water column of oceans, seas, and bodies of fresh water. Plankton abundance and distribution are strongly dependent on factors such as ambient nutrients concentrations, the physical state of the water column, and the abundance of other plankton.
Gondwana	Gondwana refers to the southern landmass formed during the Mesozoic era when continental drift split Pangaea .
Radiation	The emission of electromagnetic waves by all objects warmer than absolute zero is referred to as radiation.
Carnivore	An animal that eats a diet consisting solely of meat is referred to as a carnivore.
Gymnosperm	Gymnosperm refers to a naked-seed plant. Its seed is said to be naked because it is not

Go to **Cram101.com** for the Practice Tests for this Chapter.

enclosed in a fruit.

Extinction	In biology and ecology, extinction is the ceasing of existence of a species or group of taxa. The moment of extinction is generally considered to be the death of the last individual of that species. The death of all members of a species is extinction.
Amphibian	Amphibian is a taxon of animals that include all tetrapods (four-legged vertebrates) that do not have amniotic eggs.
Invertebrate	Invertebrate is a term coined by Jean-Baptiste Lamarck to describe any animal without a spinal column. It therefore includes all animals except vertebrates (fish, reptiles, amphibians, birds and mammals).
Herbivore	A herbivore is an animal that is adapted to eat primarily plant matter
Angiosperm	Flowering plant that produces seeds within an ovary that develops into a fruit is referred to as an angiosperm.
Vertebrate	Vertebrate is a subphylum of chordates, specifically, those with backbones or spinal columns. They started to evolve about 530 million years ago during the Cambrian explosion, which is part of the Cambrian period.
Hominid	.Hominid is any member of the biological family Hominidae, including humans, chimpanzees, gorillas, and orangutans.
Notochord	The notochord is a flexible rod-shaped body found in embryos of all chordates. It is composed of cells derived from the mesoblast and defining the primitive axis of the embryo. In lower vertebrates, it persists throughout life as the main axial support of the body, while in higher vertebrates it is replaced by the vertebral column.
Predation	Interaction in which one organism uses another, called the prey, as a food source is referred to as predation.
Population	Group of organisms of the same species occupying a certain area and sharing a common gene pool is referred to as population.

Adaptation	A biological adaptation is an anatomical structure, physiological process or behavioral trait of an organism that has evolved over a period of time by the process of natural selection such that it increases the expected long-term reproductive success of the organism.
Species	Group of similarly constructed organisms capable of interbreeding and producing fertile offspring is a species.
Population	Group of organisms of the same species occupying a certain area and sharing a common gene pool is referred to as population.
Genotype	The genotype is the specific genetic makeup (the specific genome) of an individual, usually in the form of DNA. It codes for the phenotype of that individual.
Crop	An organ, found in both earthworms and birds, in which ingested food is temporarily stored before being passed to the gizzard, where it is pulverized is the crop.
Allele	An allele is any one of a number of viable DNA codings of the same gene (sometimes the term refers to a non-gene sequence) occupying a given locus (position) on a chromosome.
Allele frequency	Allele frequency is a measure of the relative frequency of an allele on a genetic locus in a population. Usually it is expressed as a proportion or a percentage.
Gene pool	The gene pool of a species or a population is the complete set of unique alleles that would be found by inspecting the genetic material of every living member of that species or population.
Gene	Gene refers to a discrete unit of hereditary information consisting of a specific nucleotide sequence in DNA . Most of the genes of a eukaryote are located in its chromosomal DNA; a few are carried by the DNA of mitochondria and chloroplasts.
Nonrandom mating	A phenomenon in which individuals with certain genotypes sometimes mate with one another more commonly than would be expected on a random basis is nonrandom mating. Nonrandom mating does not lead to changes in allele frequencies.
Hardy-Weinberg Equilibrium	The Hardy-Weinberg equilibrium states that, under certain conditions, after one generation of random mating, the genotype frequencies at a single gene locus will become fixed at a particular equilibrium value.
Hardy-Weinberg	The Hardy-Weinberg principle states that, under certain conditions, after one generation of random mating, the genotype frequencies at a single gene locus will become fixed at a particular equilibrium value. It also specifies that those equilibrium frequencies can be represented as a simple function of the allele frequencies at that locus.
Mutation	Mutation refers to a change in the nucleotide sequence of DNA; the ultimate source of genetic diversity.
Egg	An egg is the zygote, resulting from fertilization of the ovum. It nourishes and protects the embryo.
Electrophoresis	Electrophoresis is the movement of an electrically charged substance under the influence of an electric field. This movement is due to the Lorentz force, which may be related to fundamental electrical properties of the body under study and the ambient electrical conditions.
Founder effect	Founder effect is when an isolated environment is invaded by only a few members of a species, which then multiply rapidly. In the extreme case, a single fertilized female might arrive in a new environment. It is a type of population bottleneck.
Enzyme	An enzyme is a protein that catalyzes, or speeds up, a chemical reaction. They are essential to sustain life because most chemical reactions in biological cells would occur too slowly, or would lead to different products, without them.

Go to **Cram101.com** for the Practice Tests for this Chapter.

Style	The style is a stalk connecting the stigma with the ovary below containing the transmitting tract, which facilitates the movement of the male gamete to the ovule.
Pollen	The male gametophyte in gymnosperms and angiosperms is referred to as pollen.
Heterozygous	Heterozygous means that the organism carries a different version of that gene on each of the two corresponding chromosomes.
Evolution	In biology, evolution is the process by which novel traits arise in populations and are passed on from generation to generation. Its action over large stretches of time explains the origin of new species and ultimately the vast diversity of the biological world.
Natural selection	Natural selection is the process by which biological individuals that are endowed with favorable or deleterious traits end up reproducing more or less than other individuals that do not possess such traits.
Directional selection	Directional selection occurs when natural selection favors a single allele and therefore allele frequency continuously shifts in one direction.
Disruptive selection	Disruptive selection refers to natural selection in which extreme phenotypes are favored over the average phenotype, leading to more than one distinct form.
Genetic drift	Genetic drift is the term used in population genetics to refer to the statistical drift over time of allele frequencies in a population due to random sampling effects in the formation of successive generations.
Sexual reproduction	The propagation of organisms involving the union of gametes from two parents is sexual reproduction.
Reproduction	Biological reproduction is the biological process by which new individual organisms are produced. Reproduction is a fundamental feature of all known life; each individual organism exists as the result of reproduction by an antecedent.
Polymorphism	The presence in a population of more than one allele of a gene at a frequency greater than that of newly arising mutations is referred to as polymorphism.
Herbivore	A herbivore is an animal that is adapted to eat primarily plant matter
Cell	The cell is the structural and functional unit of all living organisms, and is sometimes called the "building block of life."
Phenotype	The phenotype of an individual organism is either its total physical appearance and constitution or a specific manifestation of a trait, such as size or eye color, that varies between individuals. It is determined to some extent by genotype.
Leaf	In botany, a leaf is an above-ground plant organ specialized for photosynthesis. For this purpose, a leaf is typically flat (laminar) and thin, to expose the chloroplast containing cells (chlorenchyma tissue) to light over a broad area, and to allow light to penetrate fully into the tissues.
Larva	A free-living, sexually immature form in some animal life cycles that may differ from the adult in morphology, nutrition, and habitat is called larva.
Fossil	A preserved remnant or impression of an organism that lived in the past is referred to as fossil.

Go to **Cram101.com** for the Practice Tests for this Chapter.

Species	Group of similarly constructed organisms capable of interbreeding and producing fertile offspring is a species.
Population	Group of organisms of the same species occupying a certain area and sharing a common gene pool is referred to as population.
Gene pool	The gene pool of a species or a population is the complete set of unique alleles that would be found by inspecting the genetic material of every living member of that species or population.
Gene	Gene refers to a discrete unit of hereditary information consisting of a specific nucleotide sequence in DNA . Most of the genes of a eukaryote are located in its chromosomal DNA; a few are carried by the DNA of mitochondria and chloroplasts.
Speciation	Speciation refers to the evolutionary process by which new biological species arise. All forms of speciation have actually taken place over the course of evolution, though it still remains a subject of debate as to the relative importance of each mechanism in driving biodiversity.
Chromosome	A chromosome is, minimally, a very long, continuous piece of DNA, which contains many genes, regulatory elements and other intervening nucleotide sequences.
Allopatric speciation	Allopatric speciation, also known as geographic speciation, occurs when populations physically isolated by an extrinsic barrier evolve intrinsic (genetic) reproductive isolation such that if the barrier between the populations breaks down, individuals of the two populations can no longer interbreed.
Gene flow	The gain or loss of alleles from a population by the movement of individuals or gametes into or out of the population is called gene flow.
Pollen	The male gametophyte in gymnosperms and angiosperms is referred to as pollen.
Sympatric speciation	The formation of a new species as a result of a genetic change that produces a reproductive barrier between the changed population and the parent population; sympatric speciation occurs without a geographic barrier.
Cell division	Cell division is the process by which a cell (called the parent cell) divides into two cells (called daughter cells). Cell division is usually a small segment of a larger cell cycle. In meiosis, however, a cell is permanently transformed and cannot divide again.
Cell	The cell is the structural and functional unit of all living organisms, and is sometimes called the "building block of life."
Polyploidy	Polyploidy are cells or organisms that contain more than two copies of each of their chromosomes. Polyploids are defined relative to the behavior of their chromosomes at meiosis.
Diploid	Diploid cells have two copies (homologs) of each chromosome (both sex- and non-sex determining chromosomes), usually one from the mother and one from the father. Most somatic cells (body cells) of complex organisms are diploid.
Hybrid	Hybrid refers to the offspring of parents of two different species or of two different varieties of one species; the offspring of two parents that differ in one or more inherited traits; an individual that is heterozygous for one or more pair of genes.
Reproductive isolation	Reproductive isolation refers to the failure of organisms of one population to breed successfully with members of another; may be due to premating or postmating isolating mechanisms.
Geographic isolation	A condition in which part of the gene pool is separated by geographic barriers from the rest of the population is referred to as geographic isolation.

Go to **Cram101.com** for the Practice Tests for this Chapter.

Hybrid sterility	Hybrid sterility refers to a type of postzygotic barrier between species; the species remain isolated because hybrids fail to produce functional gametes.
Natural selection	Natural selection is the process by which biological individuals that are endowed with favorable or deleterious traits end up reproducing more or less than other individuals that do not possess such traits.
Radiation	The emission of electromagnetic waves by all objects warmer than absolute zero is referred to as radiation.
Chloroplast	A chloroplast is an organelle found in plant cells and eukaryotic algae which conduct photosynthesis. They are similar to mitochondria but are found only in plants. They are surrounded by a double membrane with an intermembrane space; they have their own DNA and are involved in energy metabolism;

Phylogeny	The science that explores the evolutionary relationships among organisms and seeks to reconstruct evolutionary history is phylogeny.
Evolution	In biology, evolution is the process by which novel traits arise in populations and are passed on from generation to generation. Its action over large stretches of time explains the origin of new species and ultimately the vast diversity of the biological world.
Species	Group of similarly constructed organisms capable of interbreeding and producing fertile offspring is a species.
Convergent evolution	An evolutionary pattern in which widely different organisms show similar characteristics because of similar ecological roles and selection pressures is called convergent evolution.
Genus	In biology, a genus is a taxonomic grouping. That is, in the classification of living organisms, a genus is considered to be distinct from other such genera. A genus has one or more species: if it has more than one species these are likely to be morphologically more similar than species belonging to different genera.
Parsimony	In scientific studies, the search for the least complex explanation for an observed phenomenon is parsimony.
Fossil	A preserved remnant or impression of an organism that lived in the past is referred to as fossil.
Vertebrate	Vertebrate is a subphylum of chordates, specifically, those with backbones or spinal columns. They started to evolve about 530 million years ago during the Cambrian explosion, which is part of the Cambrian period.
Chloroplast	A chloroplast is an organelle found in plant cells and eukaryotic algae which conduct photosynthesis. They are similar to mitochondria but are found only in plants. They are surrounded by a double membrane with an intermembrane space; they have their own DNA and are involved in energy metabolism;
Pollen	The male gametophyte in gymnosperms and angiosperms is referred to as pollen.
Flower	A flower is the reproductive structure of a flowering plant. The flower structure contains the plant's reproductive organs, and its function is to produce seeds through sexual reproduction.
Lungfish	Lungfish is a sarcopterygian fish that can breathe air (and in some species are obligate air-breathers), and have limb-like appendages instead of fins. There are six living species known; four in Africa, and one each in South America and Australia.
Biology	Biology is the branch of science dealing with the study of life. It is concerned with the characteristics, classification, and behaviors of organisms, how species come into existence, and the interactions they have with each other and with the environment.
Monophyletic	A group is monophyletic if it consists of a common ancestor and all its descendants. A taxonomic group that contain organisms but not their common ancestor is called polyphyletic, and a group that contains some but not all descendants of the most recent common ancestor is called paraphyletic.
Gymnosperm	Gymnosperm refers to a naked-seed plant. Its seed is said to be naked because it is not enclosed in a fruit.
Pedigree	A record of one's ancestors, offspring, siblings, and their offspring that may be used to determine the pattern of certain genes or disease inheritance within a family is a pedigree.

Go to **Cram101.com** for the Practice Tests for this Chapter.

Molecule	A molecule is the smallest particle of a pure chemical substance that still retains its chemical composition and properties.
Biology	Biology is the branch of science dealing with the study of life. It is concerned with the characteristics, classification, and behaviors of organisms, how species come into existence, and the interactions they have with each other and with the environment.
Nucleotide	A nucleotide is a chemical compound that consists of a heterocyclic base, a sugar, and one or more phosphate groups. In the most common nucleotides the base is a derivative of purine or pyrimidine, and the sugar is pentose - deoxyribose or ribose. They are the structural units of RNA and DNA.
Biochemistry	Biochemistry studies how complex chemical reactions give rise to life. It is a hybrid branch of chemistry which specialises in the chemical processes in living organisms.
Evolution	In biology, evolution is the process by which novel traits arise in populations and are passed on from generation to generation. Its action over large stretches of time explains the origin of new species and ultimately the vast diversity of the biological world.
Natural selection	Natural selection is the process by which biological individuals that are endowed with favorable or deleterious traits end up reproducing more or less than other individuals that do not possess such traits.
Population	Group of organisms of the same species occupying a certain area and sharing a common gene pool is referred to as population.
Protein	A protein is a complex, high-molecular-weight organic compound that consists of amino acids joined by peptide bonds. They are essential to the structure and function of all living cells and viruses. Many are enzymes or subunits of enzymes.
Amino acid	An amino acid is any molecule that contains both amino and carboxylic acid functional groups. They are the basic structural building units of proteins. They form short polymer chains called peptides or polypeptides which in turn form structures called proteins.
Acid	An acid is a water-soluble, sour-tasting chemical compound that when dissolved in water, gives a solution with a pH of less than 7.
Species	Group of similarly constructed organisms capable of interbreeding and producing fertile offspring is a species.
Gene	Gene refers to a discrete unit of hereditary information consisting of a specific nucleotide sequence in DNA . Most of the genes of a eukaryote are located in its chromosomal DNA; a few are carried by the DNA of mitochondria and chloroplasts.
Neutral mutation	Neutral mutation refers to a mutation that changes the nucleotide sequence of a gene but has little or no effect on the function of the organism.
Mutation	Mutation refers to a change in the nucleotide sequence of DNA; the ultimate source of genetic diversity.
Nucleic acid	A nucleic acid is a complex, high-molecular-weight biochemical macromolecule composed of nucleotide chains that convey genetic information. The most common are deoxyribonucleic acid (DNA) and ribonucleic acid (RNA). They are found in all living cells and viruses.
Codon	Codon refers to a three-nucleotide sequence in mRNA that specifies a particular amino acid or polypeptide termination signal; the basic unit of the genetic code.
Citric acid cycle	In aerobic organisms, the citric acid cycle is a metabolic pathway that forms part of the break down of carbohydrates, fats and proteins into carbon dioxide and water in order to generate energy. It is the second of three metabolic pathways that are involved in fuel molecule catabolism and ATP production, the other two being glycolysis and oxidative

Go to **Cram101.com** for the Practice Tests for this Chapter.

phosphorylation.

Duplication	Duplication refers to repetition of part of a chromosome resulting from fusion with a fragment from a homologous chromosome; can result from an error in meiosis or from mutagenesis.
Chromosome	A chromosome is, minimally, a very long, continuous piece of DNA, which contains many genes, regulatory elements and other intervening nucleotide sequences.
Sex chromosome	The X or Y chromosome in human beings that determines the sex of an individual. Females have two X chromosomes in diploid cells; males have an X and a Y chromosome. The sex chromosome comprises the 23rd chromosome pair in a karyotype.
Myoglobin	Myoglobin is a single-chain protein of 153 amino acids, containing a heme (iron-containing porphyrin) group in the center. With a molecular weight of 16,700 Daltons, it is the primary oxygen-carrying pigment of muscle tissues.
Homeotic gene	A homeotic gene can, when modified cause homeotic mutations or bizarre and fundamental developemental reorganizations of the body. Homeotic genes in general transform parts of the body into structures appropriate to other positions.
Homeobox	A homeobox is a DNA sequence found within genes that are involved in the regulation of development of animals, fungi and plants.
Multicellular	Multicellular organisms are those organisms consisting of more than one cell, and having differentiated cells that perform specialized functions. Most life that can be seen with the naked eye is multicellular, as are all animals (i.e. members of the kingdom Animalia) and plants (i.e. members of the kingdom Plantae).
Monomer	In chemistry, a monomer is a small molecule that may become chemically bonded to other monomers to form a polymer.
Hemoglobin	Hemoglobin is the iron-containing oxygen-transport metalloprotein in the red cells of the blood in mammals and other animals. Hemoglobin transports oxygen from the lungs to the rest of the body, such as to the muscles, where it releases the oxygen load.
Enzyme	An enzyme is a protein that catalyzes, or speeds up, a chemical reaction. They are essential to sustain life because most chemical reactions in biological cells would occur too slowly, or would lead to different products, without them.
Cell	The cell is the structural and functional unit of all living organisms, and is sometimes called the "building block of life."
Bacteria	The domain that contains procaryotic cells with primarily diacyl glycerol diesters in their membranes and with bacterial rRNA. Bacteria also is a general term for organisms that are composed of procaryotic cells and are not multicellular.
Digestion	Digestion refers to the mechanical and chemical breakdown of food into molecules small enough for the body to absorb; the second main stage of food processing, following ingestion.
Fermentation	Fermentation is the anaerobic metabolic breakdown of a nutrient molecule, such as glucose, without net oxidation. Fermentation does not release all the available energy in a molecule; it merely allows glycolysis to continue by replenishing reduced coenzymes.
Fossil	A preserved remnant or impression of an organism that lived in the past is referred to as fossil.
Genome	The genome of an organism is the whole hereditary information of an organism that is encoded in the DNA (or, for some viruses, RNA). This includes both the genes and the non-coding sequences. The genome of an organism is a complete DNA sequence of one set of chromosomes.

Prokaryote	A prokaryote is an organisms without a cell nucleus, or indeed any other membrane-bound organelles, in most cases unicellular (in rare cases, multicellular). This is in contrast to a eukaryote, organisms that have cell nuclei and may be variously unicellular or multicellular.
Ribosome	A ribosome is an organelle composed of rRNA and ribosomal proteins. It translates mRNA into a polypeptide chain (e.g., a protein). It can be thought of as a factory that builds a protein from a set of genetic instructions.
Y chromosome	Male sex chromosome that carries genes involved in sex determination is referred to as the Y chromosome. It contains the genes that cause testis development, thus determining maleness.

Element	A chemical element, often called simply element, is a chemical substance that cannot be divided or changed into other chemical substances by any ordinary chemical technique. An element is a class of substances that contain the same number of protons in all its atoms.
Genome	The genome of an organism is the whole hereditary information of an organism that is encoded in the DNA (or, for some viruses, RNA). This includes both the genes and the non-coding sequences. The genome of an organism is a complete DNA sequence of one set of chromosomes.
Ribosome	A ribosome is an organelle composed of rRNA and ribosomal proteins. It translates mRNA into a polypeptide chain (e.g., a protein). It can be thought of as a factory that builds a protein from a set of genetic instructions.
Free energy	The term thermodynamic free energy denotes the total amount of energy in a physical system which can be converted to do work.
Oxidizing atmosphere	An atmosphere that contains molecular oxygen is an oxidizing atmosphere.
Atmosphere	Earth's atmosphere is a layer of gases surrounding the planet Earth and retained by the Earth's gravity. It contains roughly 78% nitrogen and 21% oxygen, with trace amounts of other gases.
Nucleic acid	A nucleic acid is a complex, high-molecular-weight biochemical macromolecule composed of nucleotide chains that convey genetic information. The most common are deoxyribonucleic acid (DNA) and ribonucleic acid (RNA). They are found in all living cells and viruses.
Acid	An acid is a water-soluble, sour-tasting chemical compound that when dissolved in water, gives a solution with a pH of less than 7.
Mutation	Mutation refers to a change in the nucleotide sequence of DNA; the ultimate source of genetic diversity.
Metabolism	Metabolism is the biochemical modification of chemical compounds in living organisms and cells. This includes the biosynthesis of complex organic molecules (anabolism) and their breakdown (catabolism).
Oxidation	Oxidation refers to the loss of electrons from a substance involved in a redox reaction; always accompanies reduction.
Catalyst	A chemical that speeds up a reaction but is not used up in the reaction is a catalyst.
Intron	An intron is a section of DNA within a gene that does not encode part of the protein that the gene produces, and is spliced out of the mRNA that is transcribed from the gene before it is translated.
Plasma	In physics and chemistry, a plasma is an ionized gas, and is usually considered to be a distinct phase of matter. "Ionized" in this case means that at least one electron has been dissociated from a significant fraction of the molecules.
Evolution	In biology, evolution is the process by which novel traits arise in populations and are passed on from generation to generation. Its action over large stretches of time explains the origin of new species and ultimately the vast diversity of the biological world.
Molecule	A molecule is the smallest particle of a pure chemical substance that still retains its chemical composition and properties.
Bacteria	The domain that contains procaryotic cells with primarily diacyl glycerol diesters in their membranes and with bacterial rRNA. Bacteria also is a general term for organisms that are composed of procaryotic cells and are not multicellular.
Cyanobacteria	Cyanobacteria are a phylum of bacteria that obtain their energy through photosynthesis. They

Go to **Cram101.com** for the Practice Tests for this Chapter.

are often referred to as blue-green algae, even though it is now known that they are not directly related to any of the other algal groups, which are all eukaryotes.

Anaerobic   An anaerobic organism is any organism that does not require oxygen for growth.

Multicellular   Multicellular organisms are those organisms consisting of more than one cell, and having differentiated cells that perform specialized functions. Most life that can be seen with the naked eye is multicellular, as are all animals (i.e. members of the kingdom Animalia) and plants (i.e. members of the kingdom Plantae).

Go to **Cram101.com** for the Practice Tests for this Chapter.

Bacterium	Most bacterium are microscopic and unicellular, with a relatively simple cell structure lacking a cell nucleus, and organelles such as mitochondria and chloroplasts. They are the most abundant of all organisms. They are ubiquitous in soil, water, and as symbionts of other organisms.
Bacteria	The domain that contains procaryotic cells with primarily diacyl glycerol diesters in their membranes and with bacterial rRNA. Bacteria also is a general term for organisms that are composed of procaryotic cells and are not multicellular.
Species	Group of similarly constructed organisms capable of interbreeding and producing fertile offspring is a species.
Archaea	The Archaea are a major division of living organisms. Although there is still uncertainty in the exact phylogeny of the groups, Archaea, Eukaryotes and Bacteria are the fundamental classifications in what is called the three-domain system.
Biochemistry	Biochemistry studies how complex chemical reactions give rise to life. It is a hybrid branch of chemistry which specialises in the chemical processes in living organisms.
Domain	In biology, a domain is the top-level grouping of organisms in scientific classification.
Glycolysis	Glycolysis refers to the multistep chemical breakdown of a molecule of glucose into two molecules of pyruvic acid; the first stage of cellular respiration in all organisms; occurs in the cytoplasmic fluid.
Biosphere	The biosphere is that part of a planet's outer shell — including air, land, surface rocks and water — within which life occurs, and which biotic processes in turn alter or transform.
Metabolism	Metabolism is the biochemical modification of chemical compounds in living organisms and cells. This includes the biosynthesis of complex organic molecules (anabolism) and their breakdown (catabolism).
Habitat	Habitat refers to a place where an organism lives; an environmental situation in which an organism lives.
Rods	Rods, are photoreceptor cells in the retina of the eye that can function in less intense light than can the other type of photoreceptor, cone cells.
Unicellular	Unicellular organisms carry out all the functions of life. Unicellular species are those whose members consist of a single cell throughout their life cycle. This latter qualification is significant since most multicellular organisms consist of a single cell at the beginning of their life cycles.
Fission	A means of asexual reproduction whereby a parent separates into two or more genetically identical individuals of about equal size is referred to as fission.
Filament	The stamen is the male organ of a flower. Each stamen generally has a stalk called the filament, and, on top of the filament, an anther. The filament is a long chain of proteins, such as those found in hair, muscle, or in flagella.
Prokaryotic cell	A cell lacking a membrane-bounded nucleus and organelles is referred to as a prokaryotic cell.
Cell	The cell is the structural and functional unit of all living organisms, and is sometimes called the "building block of life."
Chromosome	A chromosome is, minimally, a very long, continuous piece of DNA, which contains many genes, regulatory elements and other intervening nucleotide sequences.
Organelle	In cell biology, an organelle is one of several structures with specialized functions, suspended in the cytoplasm of a eukaryotic cell.

Cytoplasm	Cytoplasm refers to everything inside a cell between the plasma membrane and the nucleus; consists of a semifluid medium and organelles.
Electron	The electron is a light fundamental subatomic particle that carries a negative electric charge. The electron is a spin-1/2 lepton, does not participate in strong interactions and has no substructure.
Cytoskeleton	Cytoskeleton refers to a meshwork of fine fibers in the cytoplasm of a eukaryotic cell; includes microfilaments, intermediate filaments, and microtubules.
Cyanobacteria	Cyanobacteria are a phylum of bacteria that obtain their energy through photosynthesis. They are often referred to as blue-green algae, even though it is now known that they are not directly related to any of the other algal groups, which are all eukaryotes.
Flagella	Flagella are whip-like organelle that many unicellular organisms, and some multicellular ones, use to move about.
Flagellum	A flagellum is a whip-like organelle that many unicellular organisms, and some multicellular ones, use to move about.
Cell wall	Cell wall refers to a protective layer external to the plasma membrane in plant cells, bacteria, fungi, and some protists; protects the cell and helps maintain its shape.
Gram stain	Gram stain refers to a stain that is selectively taken up by the cell walls of certain types of bacteria and rejected by the cell walls of others ; used to distinguish bacteria on the basis of their cell wall construction.
Microscope	A microscope is an instrument for viewing objects that are too small to be seen by the naked or unaided eye.
Plasma membrane	Membrane surrounding the cytoplasm that consists of a phospholipid bilayer with embedded proteins is referred to as plasma membrane.
Plasma	In physics and chemistry, a plasma is an ionized gas, and is usually considered to be a distinct phase of matter. "Ionized" in this case means that at least one electron has been dissociated from a significant fraction of the molecules.
Prokaryote	A prokaryote is an organisms without a cell nucleus, or indeed any other membrane-bound organelles, in most cases unicellular (in rare cases, multicellular). This is in contrast to a eukaryote, organisms that have cell nuclei and may be variously unicellular or multicellular.
Cellular respiration	Cellular respiration is the process in which the chemical bonds of energy-rich molecules such as glucose are converted into energy usable for life processes.
Respiration	Respiration is the process by which an organism obtains energy by reacting oxygen with glucose to give water, carbon dioxide and ATP (energy). Respiration takes place on a cellular level in the mitochondria of the cells and provide the cells with energy.
Chlorophyll	Chlorophyll is a green photosynthetic pigment found in plants, algae, and cyanobacteria. In plant photosynthesis incoming light is absorbed by chlorophyll and other accessory pigments in the antenna complexes of photosystem I and photosystem II.
Nitrogen fixation	Nitrogen fixation is the process by which nitrogen is taken from its relatively inert molecular form (N_2) in the atmosphere and converted into nitrogen compounds useful for other chemical processes.
Nitrogen	Nitrogen is a chemical element in the periodic table which has the symbol N and atomic number 7. Commonly a colorless, odorless, tasteless and mostly inert diatomic non-metal gas, nitrogen constitutes 78.08% percent of Earth's atmosphere and is a constituent of all living tissues. Nitrogen forms many important compounds such as amino acids, ammonia, nitric acid,

Go to **Cram101.com** for the Practice Tests for this Chapter.

and cyanides.

Fixation	Fixation in population genetics occurs when the frequency of a gene reaches 1. Fixation in biochemistry, histology, cell biology and pathology refers to the technique of preserving a specimen for microscopic study, making it intact and stable, but dead.
Nitrification	Nitrification is the biological oxidation of ammonia with oxygen into nitrite followed by the oxidation of these nitrites into nitrates. Nitrification is an important step in the nitrogen cycle in soil.
Ammonia	Ammonia is a compound of nitrogen and hydrogen with the formula NH_3. At standard temperature and pressure ammonia is a gas. It is toxic and corrosive to some materials, and has a characteristic pungent odor.
Nutrition	Nutrition refers to collectively, the processes involved in taking in, assimilating, and utilizing nutrients.
Nitrogen cycle	Nitrogen cycle refers to continuous process by which nitrogen circulates in the air, soil, water, and organisms of the biosphere.
Ion	Ion refers to an atom or molecule that has gained or lost one or more electrons, thus acquiring an electrical charge.
Cellulase	Cellulase refers to an enzyme that catalyzes the breakdown of the carbohydrate cellulose into its component glucose molecules; almost entirely restricted to microorganisms.
Host	Host is an organism that harbors a parasite, mutual partner, or commensal partner; or a cell infected by a virus.
Fever	Fever (also known as pyrexia, or a febrile response, and archaically known as ague) is a medical symptom that describes an increase in internal body temperature to levels that are above normal (37°C, 98.6°F).
Endotoxin	Endotoxin is part of the outer membrane of the cell wall of Gram-negative bacteria. It refers to the lipopolysaccharide complex associated with the outer membrane of Gram-negative bacteria.
Tetanus	Tetanus is a serious and often fatal disease caused by the neurotoxin tetanospasmin which is produced by the Gram-positive, obligate anaerobic bacterium Clostridium tetani. Tetanus also refers to a state of muscle tension.
Nucleotide	A nucleotide is a chemical compound that consists of a heterocyclic base, a sugar, and one or more phosphate groups. In the most common nucleotides the base is a derivative of purine or pyrimidine, and the sugar is pentose - deoxyribose or ribose. They are the structural units of RNA and DNA.
Gene	Gene refers to a discrete unit of hereditary information consisting of a specific nucleotide sequence in DNA . Most of the genes of a eukaryote are located in its chromosomal DNA; a few are carried by the DNA of mitochondria and chloroplasts.
Transformation	Transformation is the genetic alteration of a cell resulting from the introduction, uptake and expression of foreign genetic material (DNA or RNA).
Mutation	Mutation refers to a change in the nucleotide sequence of DNA; the ultimate source of genetic diversity.
Evolution	In biology, evolution is the process by which novel traits arise in populations and are passed on from generation to generation. Its action over large stretches of time explains the origin of new species and ultimately the vast diversity of the biological world.
Plasmid	Plasmid is a circular double-stranded DNA molecule that is separate from the chromosomal DNA.

Go to **Cram101.com** for the Practice Tests for this Chapter.

	They usually occur in bacteria and often contain genes that confer a selective advantage to the bacterium harboring them, e.g., the ability to make the bacterium antibiotic resistant.
Photosynthesis	Photosynthesis is a biochemical process in which plants, algae, and some bacteria harness the energy of light to produce food. Ultimately, nearly all living things depend on energy produced from photosynthesis for their nourishment, making it vital to life on Earth.
Spore	Spore refers to a differentiated, specialized form that can be used for dissemination, for survival of adverse conditions because of its heat and dessication resistance, and/or for reproduction. They are usually unicellular and may develop into vegetative organisms or gametes. A spore may be produced asexually or sexually and are of many types.
Reproduction	Biological reproduction is the biological process by which new individual organisms are produced. Reproduction is a fundamental feature of all known life; each individual organism exists as the result of reproduction by an antecedent.
Cell body	The part of a cell, such as a neuron, that houses the nucleus is the cell body.
Endospore	An endospore is any spore that is produced within an organism (usually a bacterium).They can usually make 2-3 of them depending on the type of bacterium.
Dormancy	Dormancy is a period of arrested plant growth. It is a survival strategy exhibited by many plant species, which enables them to survive in climates where part of the year is unsuitable for growth.
Genus	In biology, a genus is a taxonomic grouping. That is, in the classification of living organisms, a genus is considered to be distinct from other such genera. A genus has one or more species: if it has more than one species these are likely to be morphologically more similar than species belonging to different genera.
Antibiotic	Antibiotic refers to substance such as penicillin or streptomycin that is toxic to microorganisms. Usually a product of a particular microorvanism or plant.
Lipid	Lipid is one class of aliphatic hydrocarbon-containing organic compounds essential for the structure and function of living cells. They are characterized by being water-insoluble but soluble in nonpolar organic solvents.
Methanogen	Methanogen refers to type of archaea that lives in oxygen-free habitats, such as swamps, and releases methane gas. They produce methane as a metabolic by-product and are common in wetland, where they are responsible for marsh gas, and in the guts of animals such as ruminants and humans, where they are responsible for flatulence.
Extreme halophile	Microorganisms that live in unusually highly saline environments such as the Great Salt Lake or the Dead Sea are referred to as extreme halophile.
Halophile	Halophile is an extremophile that thrives in environments with very high concentrations of salt (at least 0.2 M). These are the primary inhabitants of salt lakes and inland seas where they tint the sediments bright colors.

Bacteria	The domain that contains procaryotic cells with primarily diacyl glycerol diesters in their membranes and with bacterial rRNA. Bacteria also is a general term for organisms that are composed of procaryotic cells and are not multicellular.
Amoeba	Amoeba is a genus of protozoa that moves by means of temporary projections called pseudopods, and is well-known as a representative unicellular organism. Amoeba itself is found in freshwater, typically on decaying vegetation from streams, but is not especially common in nature.
Domain	In biology, a domain is the top-level grouping of organisms in scientific classification.
Evolution	In biology, evolution is the process by which novel traits arise in populations and are passed on from generation to generation. Its action over large stretches of time explains the origin of new species and ultimately the vast diversity of the biological world.
Multicellular	Multicellular organisms are those organisms consisting of more than one cell, and having differentiated cells that perform specialized functions. Most life that can be seen with the naked eye is multicellular, as are all animals (i.e. members of the kingdom Animalia) and plants (i.e. members of the kingdom Plantae).
Unicellular	Unicellular organisms carry out all the functions of life. Unicellular species are those whose members consist of a single cell throughout their life cycle. This latter qualification is significant since most multicellular organisms consist of a single cell at the beginning of their life cycles.
Eukaryotic cell	Eukaryotic cell refers to a type of cell that has a membrane-enclosed nucleus and other membrane enclosed organelles. All organisms except bacteria and archaea are composed of eukaryotic cells.
Cell	The cell is the structural and functional unit of all living organisms, and is sometimes called the "building block of life."
Cytoskeleton	Cytoskeleton refers to a meshwork of fine fibers in the cytoplasm of a eukaryotic cell; includes microfilaments, intermediate filaments, and microtubules.
Eukaryote	A eukaryote is an organism with complex cells, in which the genetic material is organized into membrane-bound nuclei. They comprise animals, plants, and fungi—which are mostly multicellular—as well as various other groups that are collectively classified as protists.
Cyanobacteria	Cyanobacteria are a phylum of bacteria that obtain their energy through photosynthesis. They are often referred to as blue-green algae, even though it is now known that they are not directly related to any of the other algal groups, which are all eukaryotes.
Gene	Gene refers to a discrete unit of hereditary information consisting of a specific nucleotide sequence in DNA . Most of the genes of a eukaryote are located in its chromosomal DNA; a few are carried by the DNA of mitochondria and chloroplasts.
Bark	Bark is the outermost layer of stems and roots of woody plants such as trees. It overlays the wood and consists of three layers, the cork, the phloem, and the vascular cambium.
Metabolism	Metabolism is the biochemical modification of chemical compounds in living organisms and cells. This includes the biosynthesis of complex organic molecules (anabolism) and their breakdown (catabolism).
Protist	A protist is a heterogeneous group of living things, comprising those eukaryotes that are neither animals, plants, nor fungi. They are a paraphyletic grade, rather than a natural group, and do not have much in common besides a relatively simple organization
Contractile vacuole	A fluid-filled vacuole in certain protists that takes up excess water from the cytoplasm, contracts, and expels the water outside the cell through a pore in the plasma membrane is a

Go to **Cram101.com** for the Practice Tests for this Chapter.

Go to **Cram101.com** for the Practice Tests for this Chapter.
And, **NEVER** highlight a book again!

contractile vacuole.

Vacuole	A vacuole is a large membrane-bound compartment within some eukaryotic cells where they serve a variety of different functions: capturing food materials or unwanted structural debris surrounding the cell, sequestering materials that might be toxic to the cell, maintaining fluid balance (called turgor) within the cell.
Solute	Substance that is dissolved in a solvent, forming a solution is referred to as a solute.
Vesicle	In cell biology, a vesicle is a relatively small and enclosed compartment, separated from the cytosol by at least one lipid bilayer.
Plasma membrane	Membrane surrounding the cytoplasm that consists of a phospholipid bilayer with embedded proteins is referred to as plasma membrane.
Plasma	In physics and chemistry, a plasma is an ionized gas, and is usually considered to be a distinct phase of matter. "Ionized" in this case means that at least one electron has been dissociated from a significant fraction of the molecules.
Host	Host is an organism that harbors a parasite, mutual partner, or commensal partner; or a cell infected by a virus.
Binary fission	A means of asexual reproduction in which a parent organism, often a single cell, divides into two individuals of about equal size is called binary fission.
Fission	A means of asexual reproduction whereby a parent separates into two or more genetically identical individuals of about equal size is referred to as fission.
Haploid	Haploid cells bear one copy of each chromosome.
Habitat	Habitat refers to a place where an organism lives; an environmental situation in which an organism lives.
Molecular biology	Molecular biology overlaps with other areas of biology and chemistry, particularly genetics and biochemistry. Molecular biology chiefly concerns itself with understanding the interactions between the various systems of a cell, including the interrelationship of DNA, RNA and protein synthesis and learning how these interactions are regulated.
Biology	Biology is the branch of science dealing with the study of life. It is concerned with the characteristics, classification, and behaviors of organisms, how species come into existence, and the interactions they have with each other and with the environment.
Monophyletic	A group is monophyletic if it consists of a common ancestor and all its descendants. A taxonomic group that contain organisms but not their common ancestor is called polyphyletic, and a group that contains some but not all descendants of the most recent common ancestor is called paraphyletic.
Flagella	Flagella are whip-like organelle that many unicellular organisms, and some multicellular ones, use to move about.
Flagellum	A flagellum is a whip-like organelle that many unicellular organisms, and some multicellular ones, use to move about.
Species	Group of similarly constructed organisms capable of interbreeding and producing fertile offspring is a species.
Antibiotic	Antibiotic refers to substance such as penicillin or streptomycin that is toxic to microorganisms. Usually a product of a particular microorvanism or plant.
Enzyme	An enzyme is a protein that catalyzes, or speeds up, a chemical reaction. They are essential to sustain life because most chemical reactions in biological cells would occur too slowly, or would lead to different products, without them.

Go to **Cram101.com** for the Practice Tests for this Chapter.

Vector	A vector is an organism that does not cause disease itself but which spreads infection by conveying pathogens from one host to another.
Fever	Fever (also known as pyrexia, or a febrile response, and archaically known as ague) is a medical symptom that describes an increase in internal body temperature to levels that are above normal (37°C, 98.6°F).
Invertebrate	Invertebrate is a term coined by Jean-Baptiste Lamarck to describe any animal without a spinal column. It therefore includes all animals except vertebrates (fish, reptiles, amphibians, birds and mammals).
Photosynthesis	Photosynthesis is a biochemical process in which plants, algae, and some bacteria harness the energy of light to produce food. Ultimately, nearly all living things depend on energy produced from photosynthesis for their nourishment, making it vital to life on Earth.
Genus	In biology, a genus is a taxonomic grouping. That is, in the classification of living organisms, a genus is considered to be distinct from other such genera. A genus has one or more species: if it has more than one species these are likely to be morphologically more similar than species belonging to different genera.
Parasite	A parasite is an organism that spends a significant portion of its life in or on the living tissue of a host organism and which causes harm to the host without immediately killing it. They also commonly show highly specialized adaptations allowing them to exploit host resources.
Liver	The liver is an organ in vertebrates, including humans. It plays a major role in metabolism and has a number of functions in the body including drug detoxification, glycogen storage, and plasma protein synthesis. It also produces bile, which is important for digestion.
Plasmodium	In slime molds, free-living mass of cytoplasm that moves by pseudopods on a forest flloor or in a field feeding on decaying plant material by phagocytosis is called plasmodium.
Blood	Blood is a circulating tissue composed of fluid plasma and cells. The main function of blood is to supply nutrients (oxygen, glucose) and constitutional elements to tissues and to remove waste products.
Population	Group of organisms of the same species occupying a certain area and sharing a common gene pool is referred to as population.
Cilia	Numerous short, hairlike structures projecting from the cell surface that enable locomotion are cilia.
Organelle	In cell biology, an organelle is one of several structures with specialized functions, suspended in the cytoplasm of a eukaryotic cell.
Ion	Ion refers to an atom or molecule that has gained or lost one or more electrons, thus acquiring an electrical charge.
DNA replication	DNA replication is the process of copying a double-stranded DNA strand in a cell, prior to cell division. The two resulting double strands are identical (if the replication went well), and each of them consists of one original and one newly synthesized strand.
Genetic recombination	Genetic recombination refers to the production, by crossing over and/or independent assortment of chromosomes during meiosis, of offspring with allele combinations different from those in the parents.
Recombination	Genetic recombination is the transmission-genetic process by which the combinations of alleles observed at different loci in two parental individuals become shuffled in offspring individuals.
Clone	A group of genetically identical cells or organisms derived by asexual reproduction from a

Go to **Cram101.com** for the Practice Tests for this Chapter.

single parent is called a clone.

Algae	The algae consist of several different groups of living organisms that capture light energy through photosynthesis, converting inorganic substances into simple sugars with the captured energy.
Cell wall	Cell wall refers to a protective layer external to the plasma membrane in plant cells, bacteria, fungi, and some protists; protects the cell and helps maintain its shape.
Reproduction	Biological reproduction is the biological process by which new individual organisms are produced. Reproduction is a fundamental feature of all known life; each individual organism exists as the result of reproduction by an antecedent.
Zygote	Diploid cell formed by the union of sperm and egg is referred to as zygote.
Diatom	Diatom refers to a unicellular photosynthetic alga with a unique, glassy cell wall containing silica.
Filtration	Filtration involved in passive transport is the movement of water and solute molecules across the cell membrane due to hydrostatic pressure by the cardiovascular system.
Organ	Organ refers to a structure consisting of several tissues adapted as a group to perform specific functions.
Acid	An acid is a water-soluble, sour-tasting chemical compound that when dissolved in water, gives a solution with a pH of less than 7.
Diploid	Diploid cells have two copies (homologs) of each chromosome (both sex- and non-sex determining chromosomes), usually one from the mother and one from the father. Most somatic cells (body cells) of complex organisms are diploid.
Sperm	Sperm refers to the male sex cell with three distinct parts at maturity: head, middle piece, and tail.
Meiosis	In biology, meiosis is the process that transforms one diploid cell into four haploid cells in eukaryotes in order to redistribute the diploid's cell's genome. Meiosis forms the basis of sexual reproduction and can only occur in eukaryotes.
Spore	Spore refers to a differentiated, specialized form that can be used for dissemination, for survival of adverse conditions because of its heat and dessication resistance, and/or for reproduction. They are usually unicellular and may develop into vegetative organisms or gametes. A spore may be produced asexually or sexually and are of many types.
Gametophyte	Gametophyte refers to the multicellular haploid form in the life cycle of organisms undergoing alternation of generations; mitotically produces haploid gametes that unite and grow into the sporophyte generation.
Absorption	Absorption is a physical or chemical phenomenon or a process in which atoms, molecules, or ions enter some bulk phase - gas, liquid or solid material. In nutrition, amino acids are broken down through digestion, which begins in the stomach.
Decomposer	Decomposer refers to an organism that derives its energy from organic wastes and dead organisms; also called a detritivore.
Chlorophyll	Chlorophyll is a green photosynthetic pigment found in plants, algae, and cyanobacteria. In plant photosynthesis incoming light is absorbed by chlorophyll and other accessory pigments in the antenna complexes of photosystem I and photosystem II.
Haplontic life cycle	Life cycle typical of protists in which the adult is always haploid because meiosis occurs after zygote formation is referred to as haplontic life cycle.
Diplontic life	Life cycle typical of animals in which the adult is always diploid and meiosis produces the

cycle	gametes is called diplontic life cycle.
Euglenoid	A protist characterized by one or more whiplike flagella that are used for locomotion and by a photorecepto is referred to as a euglenoid. They are one of the best-known groups of flagellates, commonly found in freshwater especially when it is rich in organic materials, with a few marine and endosymbiotic members.
Microtubule	Microtubule is a protein structure found within cells, one of the components of the cytoskeleton. They have diameter of ~ 24 nm and varying length from several micrometers to possible millimeters in axons of nerve cells. They serve as structural components within cells and are involved in many cellular processes including mitosis, cytokinesis, and vesicular transport.
Radiolarian	Radiolarian are amoeboid protozoa that produce intricate mineral skeletons, typically with a central capsule dividing the cell into inner and outer portions, called endoplasm and ectoplasm. They are found as plankton throughout the ocean, and their shells are important fossils found from the Cambrian onwards.
Plankton	Plankton are drifting organisms that inhabit the water column of oceans, seas, and bodies of fresh water. Plankton abundance and distribution are strongly dependent on factors such as ambient nutrients concentrations, the physical state of the water column, and the abundance of other plankton.
Phylum	Phylum is the second highest taxonomic classification for the kingdom Animalia (animals), between kingdom level and class level.
Cytoplasm	Cytoplasm refers to everything inside a cell between the plasma membrane and the nucleus; consists of a semifluid medium and organelles.
Protein	A protein is a complex, high-molecular-weight organic compound that consists of amino acids joined by peptide bonds. They are essential to the structure and function of all living cells and viruses. Many are enzymes or subunits of enzymes.
Sporangia	Plant or fungal structures in which spores are produced are called sporangia. Sporangia occur on angiosperms, gymnosperms, ferns, fern allies, mosses, algae, and fungi.
Sporangium	A plant or fungal structure that produces spores is referred to as sporangium. They occur on angiosperms, gymnosperms, ferns, fern allies, mosses, algae, and fungi.
Substrate	A substrate is a molecule which is acted upon by an enzyme. Each enzyme recognizes only the specific substrate of the reaction it catalyzes. A surface in or on which an organism lives.

Evolution	In biology, evolution is the process by which novel traits arise in populations and are passed on from generation to generation. Its action over large stretches of time explains the origin of new species and ultimately the vast diversity of the biological world.
Algae	The algae consist of several different groups of living organisms that capture light energy through photosynthesis, converting inorganic substances into simple sugars with the captured energy.
Vascular	In botany vascular refers to tissues that contain vessels for transporting liquids. In anatomy and physiology, vascular means related to blood vessels, which are part of the Circulatory system.
Phylum	Phylum is the second highest taxonomic classification for the kingdom Animalia (animals), between kingdom level and class level.
Monophyletic	A group is monophyletic if it consists of a common ancestor and all its descendants. A taxonomic group that contain organisms but not their common ancestor is called polyphyletic, and a group that contains some but not all descendants of the most recent common ancestor is called paraphyletic.
Mitosis	Mitosis is the process by which a cell separates its duplicated genome into two identical halves. It is generally followed immediately by cytokinesis which divides the cytoplasm and cell membrane.
Diploid	Diploid cells have two copies (homologs) of each chromosome (both sex- and non-sex determining chromosomes), usually one from the mother and one from the father. Most somatic cells (body cells) of complex organisms are diploid.
Sporophyte	A sporophyte is the diploid structure or phase of life of a sexually reproducing plant. Each living cell of the sporophyte contains two complete sets of chromosomes. The sporophyte is the dominant life form in ferns, gymnosperms, and angiosperms.
Fertilization	Fertilization is fusion of gametes to form a new organism. In animals, the process involves a sperm fusing with an ovum, which eventually leads to the development of an embryo.
Cladistic analysis	Cladistic analysis refers to the study of evolutionary history; specifically, the scientific search for monophyletic taxa, taxonomic groups composed of an ancestor and all its descendants.
Sperm	Sperm refers to the male sex cell with three distinct parts at maturity: head, middle piece, and tail.
Capillary	A capillary is the smallest of a body's blood vessels, measuring 5-10 micro meters. They connect arteries and veins, and most closely interact with tissues. Their walls are composed of a single layer of cells, the endothelium. This layer is so thin that molecules such as oxygen, water and lipids can pass through them by diffusion and enter the tissues.
Tissue	Group of similar cells which perform a common function is called tissue.
Vascular tissue	Plant tissue consisting of cells joined into tubes that transport water and nutrients throughout the plant body is referred to as vascular tissue.
Fossil	A preserved remnant or impression of an organism that lived in the past is referred to as fossil.
Tracheophyte	Tracheophyte refers to a plant that has conducting vessels; a vascular plant.
Cuticle	Cuticle in animals, a tough, nonliving outer layer of the skin. In plants, a waxy coating on the surface of stems and leaves that helps retain water.
Species	Group of similarly constructed organisms capable of interbreeding and producing fertile

Go to **Cram101.com** for the Practice Tests for this Chapter.

offspring is a species.

Gametophyte	Gametophyte refers to the multicellular haploid form in the life cycle of organisms undergoing alternation of generations; mitotically produces haploid gametes that unite and grow into the sporophyte generation.
Spore	Spore refers to a differentiated, specialized form that can be used for dissemination, for survival of adverse conditions because of its heat and dessication resistance, and/or for reproduction. They are usually unicellular and may develop into vegetative organisms or gametes. A spore may be produced asexually or sexually and are of many types.
Archegonium	An archegonium is a multicellular structure or organ of the gametophyte phase of certain plants producing and containing the ovum or female gamete.
Antheridium	An antheridium is a structure or organ of the gametophyte phase of certain plants producing and containing the spermatids or male gametes.
Egg	An egg is the zygote, resulting from fertilization of the ovum. It nourishes and protects the embryo.
Zygote	Diploid cell formed by the union of sperm and egg is referred to as zygote.
Antheridia	The structures in lower plants that bear sperm are referred to as antheridia.
Capsule	A sticky layer that surrounds the bacterial cell wall, protects the cell surface, and sometimes helps glue the cell to surfaces is called the capsule. In botany, a capsule is a type of dry fruit as in the poppy, iris, foxglove, etc. as well as another term for the sporangium of mosses and hornworts.
Cell	The cell is the structural and functional unit of all living organisms, and is sometimes called the "building block of life."
Sexual reproduction	The propagation of organisms involving the union of gametes from two parents is sexual reproduction.
Reproduction	Biological reproduction is the biological process by which new individual organisms are produced. Reproduction is a fundamental feature of all known life; each individual organism exists as the result of reproduction by an antecedent.
Chloroplast	A chloroplast is an organelle found in plant cells and eukaryotic algae which conduct photosynthesis. They are similar to mitochondria but are found only in plants. They are surrounded by a double membrane with an intermembrane space; they have their own DNA and are involved in energy metabolism;
Tracheid	In flowering plants, a type of cell in xylem that has tapered ends and pits through which water and minerals flow is called a tracheid.
Sucrose	A disaccharide composed of glucose and fructose is called sucrose.
Cell division	Cell division is the process by which a cell (called the parent cell) divides into two cells (called daughter cells). Cell division is usually a small segment of a larger cell cycle. In meiosis, however, a cell is permanently transformed and cannot divide again.
Substrate	A substrate is a molecule which is acted upon by an enzyme. Each enzyme recognizes only the specific substrate of the reaction it catalyzes. A surface in or on which an organism lives.
Shoot	In botany, the shoot is one of two primary sections of a plant; the other is the root. The shoot refers to what is generally the upper portion of a plant, and consists of stems, leaves, flowers, and fruits. It is derived from the embryonic epicotyl, the portion of the embryo above the point of attachment to the seed leaves (cotyledons).
Haploid	Haploid cells bear one copy of each chromosome.

Go to **Cram101.com** for the Practice Tests for this Chapter.

Fern	A fern is any one of a group of about 20,000 species of plants classified in the Division Pteridophyta, formerly known as Filicophyta. A fern is a vascular plant that differs from the more primitive lycophytes in having true leaves (megaphylls) and from the more advanced seed plants (gymnosperms and angiosperms) in lacking seeds.
Leaf	In botany, a leaf is an above-ground plant organ specialized for photosynthesis. For this purpose, a leaf is typically flat (laminar) and thin, to expose the chloroplast containing cells (chlorenchyma tissue) to light over a broad area, and to allow light to penetrate fully into the tissues.
Organ	Organ refers to a structure consisting of several tissues adapted as a group to perform specific functions.
Sporangia	Plant or fungal structures in which spores are produced are called sporangia. Sporangia occur on angiosperms, gymnosperms, ferns, fern allies, mosses, algae, and fungi.
Element	A chemical element, often called simply element, is a chemical substance that cannot be divided or changed into other chemical substances by any ordinary chemical technique. An element is a class of substances that contain the same number of protons in all its atoms.
Whorl	A circle of leaves or of flower parts present at a single level along an axis is called whorl. Whorl(mollusc) also refers to a single, complete 360° turn in the spiral growth of a mollusc shell.
Root	In vascular plants, the root is that organ of a plant body that typically lies below the surface of the soil. However, this is not always the case, since a root can also be aerial (that is, growing above the ground) or aerating (that is, growing up above the ground or especially above water).

Go to **Cram101.com** for the Practice Tests for this Chapter.

Haploid	Haploid cells bear one copy of each chromosome.
Cone	Cone refers to a reproductive structure of gymnosperms that produces pollen in males or eggs in females.
Spore	Spore refers to a differentiated, specialized form that can be used for dissemination, for survival of adverse conditions because of its heat and dessication resistance, and/or for reproduction. They are usually unicellular and may develop into vegetative organisms or gametes. A spore may be produced asexually or sexually and are of many types.
Species	Group of similarly constructed organisms capable of interbreeding and producing fertile offspring is a species.
Angiosperm	Flowering plant that produces seeds within an ovary that develops into a fruit is referred to as an angiosperm.
Gametophyte	Gametophyte refers to the multicellular haploid form in the life cycle of organisms undergoing alternation of generations; mitotically produces haploid gametes that unite and grow into the sporophyte generation.
Sporophyte	A sporophyte is the diploid structure or phase of life of a sexually reproducing plant. Each living cell of the sporophyte contains two complete sets of chromosomes. The sporophyte is the dominant life form in ferns, gymnosperms, and angiosperms.
Insect	An arthropod that usually has three body segments , three pairs of legs, and one or two pairs of wings is called an insect. They are the largest and (on land) most widely-distributed taxon within the phylum Arthropoda. They comprise the most diverse group of animals on the earth, with around 925,000 species described
Pollen tube	A tube that grows from a pollen grain. Male reproductive cells move through the pollen tube into the ovule.
Pollen	The male gametophyte in gymnosperms and angiosperms is referred to as pollen.
Diploid	Diploid cells have two copies (homologs) of each chromosome (both sex- and non-sex determining chromosomes), usually one from the mother and one from the father. Most somatic cells (body cells) of complex organisms are diploid.
Seed coat	Protective covering for the seed, formed by the hardening of the ovule wall is a seed coat.
Multicellular	Multicellular organisms are those organisms consisting of more than one cell, and having differentiated cells that perform specialized functions. Most life that can be seen with the naked eye is multicellular, as are all animals (i.e. members of the kingdom Animalia) and plants (i.e. members of the kingdom Plantae).
Gymnosperm	Gymnosperm refers to a naked-seed plant. Its seed is said to be naked because it is not enclosed in a fruit.
Sex chromosome	The X or Y chromosome in human beings that determines the sex of an individual. Females have two X chromosomes in diploid cells; males have an X and aY chromosome. The sex chromosome comprises the 23rd chromosome pair in a karyotype.
Chromosome	A chromosome is, minimally, a very long, continuous piece of DNA, which contains many genes, regulatory elements and other intervening nucleotide sequences.
Phylum	Phylum is the second highest taxonomic classification for the kingdom Animalia (animals), between kingdom level and class level.
Desert	A desert is a landscape form or region that receives little precipitation - less than 250 mm (10 in) per year. It is a biome characterized by organisms adapted to sparse rainfall and rapid evaporation.

Go to **Cram101.com** for the Practice Tests for this Chapter.

Xylem	The nonliving portion of a plant's vascular system that provides support and conveys water and inorganic nutrients from the roots to the rest of the plant. Xylem is made up of vessel elements and/or tracheids, water-conducting cells.
Fossil	A preserved remnant or impression of an organism that lived in the past is referred to as fossil.
Integument	Integument refers to the natural outer covering layers of an animal. Develops from the ectoderm.
Pollen grain	Pollen grain in seed plants, the sperm-producing microgametophyte.
Flower	A flower is the reproductive structure of a flowering plant. The flower structure contains the plant's reproductive organs, and its function is to produce seeds through sexual reproduction.
Conifer	Conifer, a division of Pinophyta, is one of 13 or 14 division level taxa within the Kingdom Plantae. They are woody cone-bearing seed plants with vascular tissue, the great majority being trees.
Double fertilization	In flowering plants, the formation of both a zygote and a cell with a triploid nucleus, which develops into the endosperm is double fertilization.
Fertilization	Fertilization is fusion of gametes to form a new organism. In animals, the process involves a sperm fusing with an ovum, which eventually leads to the development of an embryo.
Microgametophyte	In seed plants, the gametophyte that produces sperm is referred to as microgametophyte.
Sperm	Sperm refers to the male sex cell with three distinct parts at maturity: head, middle piece, and tail.
Endosperm	Endosperm is a triploid tissue (containing three sets of chromosomes) found in the seeds of flowering plants. It provides nutrition to the developing embryo. It is mostly composed of starch, though it can also contain oils and protein.
Leaf	In botany, a leaf is an above-ground plant organ specialized for photosynthesis. For this purpose, a leaf is typically flat (laminar) and thin, to expose the chloroplast containing cells (chlorenchyma tissue) to light over a broad area, and to allow light to penetrate fully into the tissues.
Carpel	The female part of a flower, consisting of a stalk with an ovary at the base and a stigma, which traps pollen, at the tip is the carpel.
Fruit	A fruit is the ripened ovary—together with seeds—of a flowering plant. In many species, the fruit incorporates the ripened ovary and surrounding tissues.
Cell	The cell is the structural and functional unit of all living organisms, and is sometimes called the "building block of life."
Phloem	In vascular plants, phloem is the living tissue that carries organic nutrients, particularly sucrose, to all parts of the plant where needed. In trees, the phloem is underneath and difficult to distinguish from bark,
Stamen	The stamen is the male organ of a flower. Each stamen generally has a stalk called the filament, and, on top of the filament, an anther.
Pistil	Structure in a flower that consists of a stigma, a style, and an ovule-containing ovary is called pistil.
Corolla	Corolla refers to the petals of a flower, collectively. Usually, the conspicuously colored flower whorl.

Go to **Cram101.com** for the Practice Tests for this Chapter.

Calyx	Calyx is a collective term for all of the sepals, structural components of a flower. The outermost flower whorl.
Monoecious	Monoecious refers to having unisexual flowers, conifer cones, or functionally equivalent structures of both sexes appearing on the same plant.
Petal	A petal is one member or part of the corolla of a flower. It is the inner part of the perianth that comprises the sterile parts of a flower and consists of inner and outer tepals.
Style	The style is a stalk connecting the stigma with the ovary below containing the transmitting tract, which facilitates the movement of the male gamete to the ovule.
Evolution	In biology, evolution is the process by which novel traits arise in populations and are passed on from generation to generation. Its action over large stretches of time explains the origin of new species and ultimately the vast diversity of the biological world.
Ovule	An ovule is a structure found in seed plants that develops into a seed after fertilization. It consists of an embryo sac containing the female gamete, an ovum or egg cell, surrounded by nutritive tissue, the nucellus. Outside this there are one or two coverings that provide protection, developing into the testa, or seed coat, following fertilization.
Ovary	In the flowering plants, an ovary is a part of the female reproductive organ of the flower or gynoecium.
Simple fruit	Simple fruit refers to a fruit such as an apple that develops from a flower with a single carpel and ovary.
Aggregate fruit	Compound fruits may be an aggregate fruit, in which one flower contains several ovaries which each develop into a small fruit. These small fruits are joined tightly together to make a larger fruit.
Whorl	A circle of leaves or of flower parts present at a single level along an axis is called whorl. Whorl(mollusc) also refers to a single, complete 360° turn in the spiral growth of a mollusc shell.
Cotyledon	A cotyledon is a significant part of the embryo within the seed of a plant. Upon germination, the cotyledon usually becomes the embryonic first leaves of a seedling. The number of cotyledons present is one characteristic used by botanists to classify the flowering plants (angiosperms).

Go to **Cram101.com** for the Practice Tests for this Chapter.

Go to **Cram101.com** for the Practice Tests for this Chapter.
And, **NEVER** highlight a book again!

Fungus	A fungus is a eukaryotic organism that digests its food externally and absorbs the nutrient molecules into its cells.
Unicellular	Unicellular organisms carry out all the functions of life. Unicellular species are those whose members consist of a single cell throughout their life cycle. This latter qualification is significant since most multicellular organisms consist of a single cell at the beginning of their life cycles.
Spore	Spore refers to a differentiated, specialized form that can be used for dissemination, for survival of adverse conditions because of its heat and dessication resistance, and/or for reproduction. They are usually unicellular and may develop into vegetative organisms or gametes. A spore may be produced asexually or sexually and are of many types.
Protist	A protist is a heterogeneous group of living things, comprising those eukaryotes that are neither animals, plants, nor fungi. They are a paraphyletic grade, rather than a natural group, and do not have much in common besides a relatively simple organization
Cellulose	A large polysaccharide composed of many glucose monomers linked into cable-like fibrils that provide structural support in plant cell walls is referred to as cellulose.
Taxonomy	Initially taxonomy was only the science of classifying living organisms, but later the word was applied in a wider sense, and may also refer to either a classification of things, or the principles underlying the classification. Almost anything, animate objects, inanimate objects, places, and events, may be classified according to some taxonomic scheme.
Phylum	Phylum is the second highest taxonomic classification for the kingdom Animalia (animals), between kingdom level and class level.
Substrate	A substrate is a molecule which is acted upon by an enzyme. Each enzyme recognizes only the specific substrate of the reaction it catalyzes. A surface in or on which an organism lives.
Mycelium	Mycelium is the vegetative part of a fungus consisting of a mass of branching threadlike hyphae that exists below the ground or within another substrate.
Hypertonic	A hypertonic cell environment has a higher concentration of solutes than in cytoplasm. In a hypertonic environment, osmosis causes water to flow out of the cell. If enough water is removed in this way, the cytoplasm will have such a small concentration of water that the cell has difficulty functioning.
Bacteria	The domain that contains procaryotic cells with primarily diacyl glycerol diesters in their membranes and with bacterial rRNA. Bacteria also is a general term for organisms that are composed of procaryotic cells and are not multicellular.
Decomposer	Decomposer refers to an organism that derives its energy from organic wastes and dead organisms; also called a detritivore.
Ion	Ion refers to an atom or molecule that has gained or lost one or more electrons, thus acquiring an electrical charge.
Vitamin	A Vitamin is an organic molecule required by a living organism in minute amounts for proper health. An organism deprived of all sources of a particular vitamin will eventually suffer from disease symptoms specific to that vitamin.
Adaptation	A biological adaptation is an anatomical structure, physiological process or behavioral trait of an organism that has evolved over a period of time by the process of natural selection such that it increases the expected long-term reproductive success of the organism.
Reproduction	Biological reproduction is the biological process by which new individual organisms are produced. Reproduction is a fundamental feature of all known life; each individual organism exists as the result of reproduction by an antecedent.

Go to **Cram101.com** for the Practice Tests for this Chapter.

Zygote	Diploid cell formed by the union of sperm and egg is referred to as zygote.
Gamete	A gamete is a specialized germ cell that unites with another gamete during fertilization in organisms that reproduce sexually. They are haploid cells; that is, they contain one complete set of chromosomes. When they unite they form a zygote—a cell having two complete sets of chromosomes and therefore diploid.
Crop	An organ, found in both earthworms and birds, in which ingested food is temporarily stored before being passed to the gizzard, where it is pulverized is the crop.
Genus	In biology, a genus is a taxonomic grouping. That is, in the classification of living organisms, a genus is considered to be distinct from other such genera. A genus has one or more species: if it has more than one species these are likely to be morphologically more similar than species belonging to different genera.
Haploid	Haploid cells bear one copy of each chromosome.
Pigment	Pigment is any material resulting in color in plant or animal cells which is the result of selective absorption.
Cytokinesis	The division of the cytoplasm to form two separate daughter cells. Cytokinesis usually occurs during telophase of mitosis, and the two processes make up the mitotic phase of the cell cycle.
Diploid	Diploid cells have two copies (homologs) of each chromosome (both sex- and non-sex determining chromosomes), usually one from the mother and one from the father. Most somatic cells (body cells) of complex organisms are diploid.
Flagella	Flagella are whip-like organelle that many unicellular organisms, and some multicellular ones, use to move about.
Evolution	In biology, evolution is the process by which novel traits arise in populations and are passed on from generation to generation. Its action over large stretches of time explains the origin of new species and ultimately the vast diversity of the biological world.
Fruiting body	In fungi, the fruiting body (also known as sporocarp) is a multicellular structure on which spore-producing structures, such as basidia or asci, are borne. The fruiting body is part of the sexual phase of a fungal life cycle, with the rest of the life cycle being characterized by vegetative mycelial growth.
Species	Group of similarly constructed organisms capable of interbreeding and producing fertile offspring is a species.
Sporangium	A plant or fungal structure that produces spores is referred to as sporangium. They occur on angiosperms, gymnosperms, ferns, fern allies, mosses, algae, and fungi.
Sporangia	Plant or fungal structures in which spores are produced are called sporangia. Sporangia occur on angiosperms, gymnosperms, ferns, fern allies, mosses, algae, and fungi.
Hypha	A hypha is a long, branching filament found primarily in fungi, but also in fungus-like bacteria such as Actinomyces and Streptomyces.
Septum	A septum, in general, is a wall separating two cavities or two spaces containing a less dense material. The muscle wall that divides the heart chambers.
Meiosis	In biology, meiosis is the process that transforms one diploid cell into four haploid cells in eukaryotes in order to redistribute the diploid's cell's genome. Meiosis forms the basis of sexual reproduction and can only occur in eukaryotes.
Basidium	Basidium refers to a diploid cell, typically club-shaped, formed by members of the fungal division Basidiomycota; produces basidiospores by meiosis.

Dikaryotic phase	Dikaryotic phase refers to a series in the life cycle of many fungi in which cells contain two nuclei.
Sexual reproduction	The propagation of organisms involving the union of gametes from two parents is sexual reproduction.
Root hair	Root hair refers to an outgrowth of an epidermal cell on a root, which increases the root's absorptive surface area.
Root	In vascular plants, the root is that organ of a plant body that typically lies below the surface of the soil. However, this is not always the case, since a root can also be aerial (that is, growing above the ground) or aerating (that is, growing up above the ground or especially above water).
Lichen	Lichen is a symbiotic organism in a mutualistic association between a fungus and an alga or between a fungus and a cyanobacterium.
Pollution	Any environmental change that adversely affects the lives and health of living things is referred to as pollution.
Bark	Bark is the outermost layer of stems and roots of woody plants such as trees. It overlays the wood and consists of three layers, the cork, the phloem, and the vascular cambium.
Photosynthesis	Photosynthesis is a biochemical process in which plants, algae, and some bacteria harness the energy of light to produce food. Ultimately, nearly all living things depend on energy produced from photosynthesis for their nourishment, making it vital to life on Earth.

Vertebrate	Vertebrate is a subphylum of chordates, specifically, those with backbones or spinal columns. They started to evolve about 530 million years ago during the Cambrian explosion, which is part of the Cambrian period.
Gene	Gene refers to a discrete unit of hereditary information consisting of a specific nucleotide sequence in DNA . Most of the genes of a eukaryote are located in its chromosomal DNA; a few are carried by the DNA of mitochondria and chloroplasts.
Ventral	The surface or side of the body normally oriented upwards, away from the pull of gravity, is the dorsal side; the opposite side, typically the one closest to the ground when walking on all legs, swimming or flying, is the ventral side.
Cleavage	Cleavage refers to cytokinesis in animal cells and in some protists, characterized by pinching in of the plasma membrane.
Cell	The cell is the structural and functional unit of all living organisms, and is sometimes called the "building block of life."
Body cavity	A fluid-containing space between the digestive tract and the body wall is referred to as body cavity.
Coelom	A fluid filled body cavity with a complete lining called peritoneum derived from mesoderm is called a coelom. Organs formed inside a coelom can freely move, grow, and develop independently of the body wall while fluid cushions and protects them from shocks.
Organ	Organ refers to a structure consisting of several tissues adapted as a group to perform specific functions.
Radial symmetry	Organisms with radial symmetry have body parts arranged in a regular, repeating pattern around a central axis (like a wagon wheel) or are completely symmetrical about a central axis (like a dinner plate). These organisms resemble a pie where several cutting planes produce roughly identical pieces. An organism with radial symmetry exhibits no left or right sides. They have a top and a bottom only.
Symmetry	Symmetry in biology is the balanced distribution of duplicate body parts or shapes.
Cnidaria	Cnidaria is a phylum containing some 10,000 species of relatively simple animals found exclusively in aquatic environments (most species are marine). They get their name from cnidocytes, which are specialized cells that carry stinging organelles.
Cephalization	Cephalization is the evolutionary process by which sensory organs move to the head region. This process is also tied to the development of an anterior brain in the vertebrates from the notochord.
Skeleton	In biology, the skeleton or skeletal system is the biological system providing physical support in living organisms.
Substrate	A substrate is a molecule which is acted upon by an enzyme. Each enzyme recognizes only the specific substrate of the reaction it catalyzes. A surface in or on which an organism lives.
Protist	A protist is a heterogeneous group of living things, comprising those eukaryotes that are neither animals, plants, nor fungi. They are a paraphyletic grade, rather than a natural group, and do not have much in common besides a relatively simple organization
Sponge	An invertebrates that consist of a complex aggregation of cells, including collar cells, and has a skeleton of fibers and/or spicules is a sponge. They are primitive, sessile, mostly marine, waterdwelling filter feeders that pump water through their matrix to filter out particulates of food matter.
Flagellum	A flagellum is a whip-like organelle that many unicellular organisms, and some multicellular ones, use to move about.

Go to **Cram101.com** for the Practice Tests for this Chapter.

Flagella	Flagella are whip-like organelle that many unicellular organisms, and some multicellular ones, use to move about.
Epidermis	Epidermis is the outermost layer of the skin. It forms the waterproof, protective wrap over the body's surface and is made up of stratified squamous epithelium with an underlying basement membrane. It contains no blood vessels, and is nourished by diffusion from the dermis. In plants, the outermost layer of cells covering the leaves and young parts of a plant is the epidermis.
Species	Group of similarly constructed organisms capable of interbreeding and producing fertile offspring is a species.
Sperm	Sperm refers to the male sex cell with three distinct parts at maturity: head, middle piece, and tail.
Reproduction	Biological reproduction is the biological process by which new individual organisms are produced. Reproduction is a fundamental feature of all known life; each individual organism exists as the result of reproduction by an antecedent.
Microscope	A microscope is an instrument for viewing objects that are too small to be seen by the naked or unaided eye.
Muscle	Muscle is a contractile form of tissue. It is one of the four major tissue types, the other three being epithelium, connective tissue and nervous tissue. Muscle contraction is used to move parts of the body, as well as to move substances within the body.
Sessile	In botany, sessile means without a stalk, as in flowers or leaves that grow directly from the stem. In zoology, sessile organisms are those which are not able to move about. They are usually permanently attached to a solid substrate of some kind, such as a rock, or the hull of a ship in the case of barnacles.
Egg	An egg is the zygote, resulting from fertilization of the ovum. It nourishes and protects the embryo.
Polyp	Small, abnormal growth that arises from the epithelial lining is referred to as polyp. In zoology, a polyp is one of two forms of individuals found in many species of cnidarians. They are approximately cylindrical, elongated on the axis of the body.
Mesoglea	Mesoglea is the clear, inert, jellylike substance that makes up most of the bodies of jellyfish, comb jellies and certain other primitive sea creatures. It acts as the creatures' structural support in water, as they lack bones, cartilage or other more common means of support.
Gastrovascular cavity	Gastrovascular cavity refers to a digestive compartment with a single opening, the mouth; may function in circulation, body support, waste disposal, and gas exchange, as well as digestion.
Medusa	Medusa is a form of cnidarian in which the body is shortened on its principal axis and broadened, sometimes greatly, in contrast with the hydroid or polyp.
Matrix	In biology, matrix (plural: matrices) is the material between animal or plant cells, the material (or tissue) in which more specialized structures are embedded, and a specific part of the mitochondrion that is the site of oxidation of organic molecules.
Coral	Group of cnidarians having a calcium carbonate skeleton that participate in the formation of reefs is coral.
Coral reef	A coral reef is a type of biotic reef that develops in tropical waters by the growth of coralline algae, hermatypic corals, and other marine organisms.
Global warming	A slow but steady rise in Earth's surface temperature, caused by increasing concentrations of

Go to **Cram101.com** for the Practice Tests for this Chapter.

greenhouse gases in the atmosphere is referred to as global warming.

Algae	The algae consist of several different groups of living organisms that capture light energy through photosynthesis, converting inorganic substances into simple sugars with the captured energy.
Homeobox	A homeobox is a DNA sequence found within genes that are involved in the regulation of development of animals, fungi and plants.
Antagonistic muscles	Antagonistic muscles refers to a pair of muscles, one of which contracts and in so doing extends the other; an arrangement that makes possible movement of the skeleton at joints.
Phylum	Phylum is the second highest taxonomic classification for the kingdom Animalia (animals), between kingdom level and class level.
Absorption	Absorption is a physical or chemical phenomenon or a process in which atoms, molecules, or ions enter some bulk phase - gas, liquid or solid material. In nutrition, amino acids are broken down through digestion, which begins in the stomach.
Cilia	Numerous short, hairlike structures projecting from the cell surface that enable locomotion are cilia.
Genus	In biology, a genus is a taxonomic grouping. That is, in the classification of living organisms, a genus is considered to be distinct from other such genera. A genus has one or more species: if it has more than one species these are likely to be morphologically more similar than species belonging to different genera.
Host	Host is an organism that harbors a parasite, mutual partner, or commensal partner; or a cell infected by a virus.
Ciliate	Ciliate is one of the most important groups of protists, common almost everywhere there is water. They are a type of protozoan that moves by means of cilia.
Anus	In anatomy, the anus is the external opening of the rectum. Closure is controlled by sphincter muscles. Feces are expelled from the body through the anus during the act of defecation, which is the primary function of the anus.
Tissue	Group of similar cells which perform a common function is called tissue.
Asexual reproduction	Asexual reproduction refers to reproduction that does not involve the fusion of haploid sex cells. The parent body may divide and new parts regenerate, or a new, smaller individual may form as an attachment to the parent, to drop off when complete.
Bivalve	Bivalve refers to a member of a group of mollusks that includes clams, mussels, scallops, and oysters. They typically have two-part shells, with both parts being more or less symmetrical.
Ligament	A ligament is a short band of tough fibrous connective tissue composed mainly of long, stringy collagen fibres. They connect bones to other bones to form a joint. (They do not connect muscles to bones.)
Plankton	Plankton are drifting organisms that inhabit the water column of oceans, seas, and bodies of fresh water. Plankton abundance and distribution are strongly dependent on factors such as ambient nutrients concentrations, the physical state of the water column, and the abundance of other plankton.
Fossil	A preserved remnant or impression of an organism that lived in the past is referred to as fossil.
Annelid	Annelid refers to member of a phylum of invertebrates that contains segmented worms, such as the earthworm and the clam worm.
Blood	Blood is a circulating tissue composed of fluid plasma and cells. The main function of blood

Go to **Cram101.com** for the Practice Tests for this Chapter.

is to supply nutrients (oxygen, glucose) and constitutional elements to tissues and to remove waste products.

Larva	A free-living, sexually immature form in some animal life cycles that may differ from the adult in morphology, nutrition, and habitat is called larva.
Leech	The leech is a annelid comprising the subclass Hirudinea. There are freshwater, terrestrial and marine leeches. Like their near relatives, the Oligochaeta, they share the presence of a clitellum.Like earthworms, leeches are hermaphrodites.
Mantle	The mantle is an organ found in mollusks. It is the dorsal body wall covering the main body, or visceral mass. The epidermis of this organ secretes calcium carbonate to create a shell.
Radula	A toothed, grasping organ found in many mollusks, used to scrape up or shred food is referred to as a radula.
Herbivore	A herbivore is an animal that is adapted to eat primarily plant matter
Cephalopod	Cephalopod refers to a member of a group of mollusks that includes squids and octopuses, characterized by bilateral body symmetry.
Evolution	In biology, evolution is the process by which novel traits arise in populations and are passed on from generation to generation. Its action over large stretches of time explains the origin of new species and ultimately the vast diversity of the biological world.
Multicellular	Multicellular organisms are those organisms consisting of more than one cell, and having differentiated cells that perform specialized functions. Most life that can be seen with the naked eye is multicellular, as are all animals (i.e. members of the kingdom Animalia) and plants (i.e. members of the kingdom Plantae).
Acoelomate	Acoelomate animals, like flatworms, have no body cavity at all. Mesodermal tissues between the gut and body wall hold their organs in place.
Monophyletic	A group is monophyletic if it consists of a common ancestor and all its descendants. A taxonomic group that contain organisms but not their common ancestor is called polyphyletic, and a group that contains some but not all descendants of the most recent common ancestor is called paraphyletic.
Dorsal	In anatomy, the dorsal is the side in which the backbone is located. This is usually the top of an animal, although in humans it refers to the back.
Clade	A clade is a branch in a cladogram, which is a diagram in the form of a tree resulting from a cladistic analysis. Also refers to a group of organisms which share a common ancestor and which includes the ancestor and all the descendents of that ancestor.

Exoskeleton	An exoskeleton is an external anatomical feature that supports and protects an animal's body. Many invertebrate animals such as insects, crustaceans and shellfish have an exoskeleton.
Molting	In animals, molting is the routine shedding off old feathers in birds, or of old skin in reptiles, or of old hairs in mammals. In arthropods, such as insects, arachnids and crustaceans, molting describes the shedding of its exoskeleton (which is often called its shell), typically to let it grow.
Homeobox	A homeobox is a DNA sequence found within genes that are involved in the regulation of development of animals, fungi and plants.
Cilia	Numerous short, hairlike structures projecting from the cell surface that enable locomotion are cilia.
Species	Group of similarly constructed organisms capable of interbreeding and producing fertile offspring is a species.
Algae	The algae consist of several different groups of living organisms that capture light energy through photosynthesis, converting inorganic substances into simple sugars with the captured energy.
Cuticle	Cuticle in animals, a tough, nonliving outer layer of the skin. In plants, a waxy coating on the surface of stems and leaves that helps retain water.
Parasite	A parasite is an organism that spends a significant portion of its life in or on the living tissue of a host organism and which causes harm to the host without immediately killing it. They also commonly show highly specialized adaptations allowing them to exploit host resources.
Organ	Organ refers to a structure consisting of several tissues adapted as a group to perform specific functions.
Topsoil	Topsoil is the uppermost layer of soil, usually the top 15-20 cm. It has the highest concentration of organic matter and microorganisms, and is where most of the Earth's biological soil activity occurs.
Mammal	Homeothermic vertebrate characterized especially by the presence of hair and mammary glands is a mammal.
Segmentation	Segmentation in biology refers to the division of some metazoan bodies and plant body plans into a series of semi-repetitive segments, and the question of the benefits and costs of doing so.
Protein	A protein is a complex, high-molecular-weight organic compound that consists of amino acids joined by peptide bonds. They are essential to the structure and function of all living cells and viruses. Many are enzymes or subunits of enzymes.
Sessile	In botany, sessile means without a stalk, as in flowers or leaves that grow directly from the stem. In zoology, sessile organisms are those which are not able to move about. They are usually permanently attached to a solid substrate of some kind, such as a rock, or the hull of a ship in the case of barnacles.
Crustacean	Member of a group of marine arthropods that contains among others shrimps, crabs, crayfish, and lobsters is called crustacean.
Larva	A free-living, sexually immature form in some animal life cycles that may differ from the adult in morphology, nutrition, and habitat is called larva.
Thorax	The thorax is a division of an animal's body that lies between the head and the abdomen. In humans, the thorax is the region of the body that extends from the neck to the diaphragm, not including the upper limbs.

Insect	An arthropod that usually has three body segments , three pairs of legs, and one or two pairs of wings is called an insect. They are the largest and (on land) most widely-distributed taxon within the phylum Arthropoda. They comprise the most diverse group of animals on the earth, with around 925,000 species described
Complete metamorphosis	Complete metamorphosis is a process of development in which the immature form looks and acts differently from the adult; the stages of development are egg, larva, pupa, and adult.
Metamorphosis	Metamorphosis is a process in biology by which an individual physically develops after birth or hatching, and involves significant change in form as well as growth and differentiation.
Evolution	In biology, evolution is the process by which novel traits arise in populations and are passed on from generation to generation. Its action over large stretches of time explains the origin of new species and ultimately the vast diversity of the biological world.
Protostome	A protostome is a group of coelomate animals in which the first embryonic opening is associated with the mouth. In protostome development, the mouth forms at the site of the blastopore, and the anus forms as a second opening.
Fertilization	Fertilization is fusion of gametes to form a new organism. In animals, the process involves a sperm fusing with an ovum, which eventually leads to the development of an embryo.
Reproduction	Biological reproduction is the biological process by which new individual organisms are produced. Reproduction is a fundamental feature of all known life; each individual organism exists as the result of reproduction by an antecedent.
Nutrition	Nutrition refers to collectively, the processes involved in taking in, assimilating, and utilizing nutrients.
Extinction	In biology and ecology, extinction is the ceasing of existence of a species or group of taxa. The moment of extinction is generally considered to be the death of the last individual of that species.The death of all members of a species is extinction.
Body cavity	A fluid-containing space between the digestive tract and the body wall is referred to as body cavity.

Species	Group of similarly constructed organisms capable of interbreeding and producing fertile offspring is a species.
Color vision	Ability to detect the color of an object, dependent on three kinds of cone cells is called color vision.
Vertebrate	Vertebrate is a subphylum of chordates, specifically, those with backbones or spinal columns. They started to evolve about 530 million years ago during the Cambrian explosion, which is part of the Cambrian period.
Evolution	In biology, evolution is the process by which novel traits arise in populations and are passed on from generation to generation. Its action over large stretches of time explains the origin of new species and ultimately the vast diversity of the biological world.
Sieve plate	Perforated wall area in a sieve-tube element through which strands connecting sieve-tube protoplasts pass is a sieve plate.
Echinoderm	Echinoderm is a phylum of marine animals found in the ocean at all depths. This phylum dates back to the lower Cambrian period and represents about 7000 living species and 13000 extinct ones. 6 classes made it to the Modern Era.
Digestive system	The organ system that ingests food, breaks it down into smaller chemical units, and absorbs the nutrient molecules is referred to as the digestive system.
Algae	The algae consist of several different groups of living organisms that capture light energy through photosynthesis, converting inorganic substances into simple sugars with the captured energy.
Stomach	The stomach is an organ in the alimentary canal used to digest food. It's primary function is not the absorption of nutrients from digested food; rather, the main job of the stomach is to break down large food molecules into smaller ones, so that they can be absorbed into the blood more easily.
Esophagus	The esophagus, or gullet is the muscular tube in vertebrates through which ingested food passes from the mouth area to the stomach. Food is passed through the esophagus by using the process of peristalsis.
Notochord	The notochord is a flexible rod-shaped body found in embryos of all chordates. It is composed of cells derived from the mesoblast and defining the primitive axis of the embryo. In lower vertebrates, it persists throughout life as the main axial support of the body, while in higher vertebrates it is replaced by the vertebral column.
Larva	A free-living, sexually immature form in some animal life cycles that may differ from the adult in morphology, nutrition, and habitat is called larva.
Nerve	A nerve is an enclosed, cable-like bundle of nerve fibers or axons, which includes the glia that ensheath the axons in myelin.
Protein	A protein is a complex, high-molecular-weight organic compound that consists of amino acids joined by peptide bonds. They are essential to the structure and function of all living cells and viruses. Many are enzymes or subunits of enzymes.
Salt	Salt is a term used for ionic compounds composed of positively charged cations and negatively charged anions, so that the product is neutral and without a net charge.
Digestion	Digestion refers to the mechanical and chemical breakdown of food into molecules small enough for the body to absorb; the second main stage of food processing, following ingestion.
Cartilaginous fish	A fish that has a flexible skeleton made of cartilage is a cartilaginous fish.

Go to **Cram101.com** for the Practice Tests for this Chapter.

Dorsal	In anatomy, the dorsal is the side in which the backbone is located. This is usually the top of an animal, although in humans it refers to the back.
Plankton	Plankton are drifting organisms that inhabit the water column of oceans, seas, and bodies of fresh water. Plankton abundance and distribution are strongly dependent on factors such as ambient nutrients concentrations, the physical state of the water column, and the abundance of other plankton.
Filter feeder	A Filter feeder is an animal that feeds by straining suspended matter and food particles from water, typically by passing the water over a specialized structure, such as the baleen of baleen whales.
Invertebrate	Invertebrate is a term coined by Jean-Baptiste Lamarck to describe any animal without a spinal column. It therefore includes all animals except vertebrates (fish, reptiles, amphibians, birds and mammals).
Lungfish	Lungfish is a sarcopterygian fish that can breathe air (and in some species are obligate air-breathers), and have limb-like appendages instead of fins. There are six living species known; four in Africa, and one each in South America and Australia.
Predator	A predator is an animal or other organism that hunts and kills other organisms for food in an act called predation.
Bony fish	Osteichthyes are the bony fish, a group paraphyletic to the land vertebrates, which are sometimes included. Most belong to the Actinopterygii.
Amphibian	Amphibian is a taxon of animals that include all tetrapods (four-legged vertebrates) that do not have amniotic eggs.
Radiation	The emission of electromagnetic waves by all objects warmer than absolute zero is referred to as radiation.
Skin	Skin is an organ of the integumentary system composed of a layer of tissues that protect underlying muscles and organs.
Reptile	Member of a class of terrestrial vertebrates with internal fertilization, scaly skin, and an egg with a leathery shell is called reptile.
Herbivore	A herbivore is an animal that is adapted to eat primarily plant matter
Carnivore	An animal that eats a diet consisting solely of meat is referred to as a carnivore.
Ventral	The surface or side of the body normally oriented upwards, away from the pull of gravity, is the dorsal side; the opposite side, typically the one closest to the ground when walking on all legs, swimming or flying, is the ventral side.
Fossil	A preserved remnant or impression of an organism that lived in the past is referred to as fossil.
Brain	The part of the central nervous system involved in regulating and controlling body activity and interpreting information from the senses transmitted through the nervous system is referred to as the brain.
Marsupial	A marsupial is a mammal in which the female typically has a pouch in which it rears its young through early infancy. They differ from placental mammals (Placentalia) in their reproductive traits.
Primate	A primate is any member of the biological group that contains all lemurs, monkeys, apes, and humans.
Ecology	Ecology is the scientific study of the distribution and abundance of living organisms and how these properties are affected by interactions between the organisms and their environment.

Go to **Cram101.com** for the Practice Tests for this Chapter.

Genus	In biology, a genus is a taxonomic grouping. That is, in the classification of living organisms, a genus is considered to be distinct from other such genera. A genus has one or more species: if it has more than one species these are likely to be morphologically more similar than species belonging to different genera.
Homo sapiens	Homo sapiens are bipedal primates of the superfamily Hominoidea, together with the other apes—chimpanzees, gorillas, orangutans, and gibbons. They are the dominant sentient species on planet Earth.
Homo sapien	Humans, or human beings, are classified as bipedal primates belonging to the mammalian species Homo sapien. Humans have a highly developed brain capable of abstract reasoning, language, and introspection.
Population	Group of organisms of the same species occupying a certain area and sharing a common gene pool is referred to as population.

Go to **Cram101.com** for the Practice Tests for this Chapter.

Species	Group of similarly constructed organisms capable of interbreeding and producing fertile offspring is a species.
Shoot system	Shoot system refers to aboveground portion of a plant consisting of the stem, leaves, and flowers.
Shoot	In botany, the shoot is one of two primary sections of a plant; the other is the root. The shoot refers to what is generally the upper portion of a plant, and consists of stems, leaves, flowers, and fruits. It is derived from the embryonic epicotyl, the portion of the embryo above the point of attachment to the seed leaves (cotyledons).
Taproot	A root system common to dicots consisting of one large, vertical root that produces many smaller lateral, or branch, roots is a taproot.
Fibrous root system	A fibrous root system is the opposite of a tap root system. It is usually formed by thin, moderately branching roots growing from the stem. A fibrous root system is universal in monocotyledonous plants and ferns, and is also common in dicotyledonous plants..
Root	In vascular plants, the root is that organ of a plant body that typically lies below the surface of the soil. However, this is not always the case, since a root can also be aerial (that is, growing above the ground) or aerating (that is, growing up above the ground or especially above water).
Stem	Stem refers to that part of a plant's shoot system that supports the leaves and reproductive structures.
Gymnosperm	Gymnosperm refers to a naked-seed plant. Its seed is said to be naked because it is not enclosed in a fruit.
Leaf	In botany, a leaf is an above-ground plant organ specialized for photosynthesis. For this purpose, a leaf is typically flat (laminar) and thin, to expose the chloroplast containing cells (chlorenchyma tissue) to light over a broad area, and to allow light to penetrate fully into the tissues.
Tissue	Group of similar cells which perform a common function is called tissue.
Organelle	In cell biology, an organelle is one of several structures with specialized functions, suspended in the cytoplasm of a eukaryotic cell.
Cell	The cell is the structural and functional unit of all living organisms, and is sometimes called the "building block of life."
Cytokinesis	The division of the cytoplasm to form two separate daughter cells. Cytokinesis usually occurs during telophase of mitosis, and the two processes make up the mitotic phase of the cell cycle.
Cell wall	Cell wall refers to a protective layer external to the plasma membrane in plant cells, bacteria, fungi, and some protists; protects the cell and helps maintain its shape.
Cytoplasm	Cytoplasm refers to everything inside a cell between the plasma membrane and the nucleus; consists of a semifluid medium and organelles.
Plasmodesma	The connecting strands of protoplasm between the cytoplasm of adjacent cells which forms canals through the cell walls is plasmodesma. It may contain a desmotubule which links the endoplasmic reticulum of the adjacent cells.
Middle lamella	A thin layer of sticky polysaccharides, such as pectin, and other carbohydrates that separates and holds together the primary cell walls of adjacent plant cells is called middle lamella.
Lamella	Lamella refers to each of the many thin plates that make up the gill filaments of fish gills

or cell structures associated with cellular motility.

Parenchyma	The parenchyma are the functional parts of an organ in the body (i.e. the nephrons of the kidney, the alveoli of the lungs). In plants parenchyma cells are thin-walled cells of the ground tissue that make up the bulk of most nonwoody structures, although sometimes their cell walls can be lignified.
Collenchyma	Plant tissue composed of cells with unevenly thickened walls is a collenchyma. They provide structural support, particularly in growing shoots and leaves.
Bark	Bark is the outermost layer of stems and roots of woody plants such as trees. It overlays the wood and consists of three layers, the cork, the phloem, and the vascular cambium.
Xylem	The nonliving portion of a plant's vascular system that provides support and conveys water and inorganic nutrients from the roots to the rest of the plant. Xylem is made up of vessel elements and/or tracheids, water-conducting cells.
Element	A chemical element, often called simply element, is a chemical substance that cannot be divided or changed into other chemical substances by any ordinary chemical technique. An element is a class of substances that contain the same number of protons in all its atoms.
Angiosperm	Flowering plant that produces seeds within an ovary that develops into a fruit is referred to as an angiosperm.
Phloem	In vascular plants, phloem is the living tissue that carries organic nutrients, particularly sucrose, to all parts of the plant where needed. In trees, the phloem is underneath and difficult to distinguish from bark,
Sieve tube	In phloem, a single strand of sievetube elements that transports sugar solutions is called sieve tube.
Nucleus	In cell biology, the nucleus is found in all eukaryotic cells that contains most of the cell's genetic material. The nucleus has two primary functions: to control chemical reactions within the cytoplasm and to store information needed for cellular division.
Vascular tissue	Plant tissue consisting of cells joined into tubes that transport water and nutrients throughout the plant body is referred to as vascular tissue.
Vascular	In botany vascular refers to tissues that contain vessels for transporting liquids. In anatomy and physiology, vascular means related to blood vessels, which are part of the Circulatory system.
Node	In botany, a node is the place on a stem where a leaf is attached.
Dermal tissue system	Dermal tissue system refers to a plant tissue system that makes up the outer covering of the plant body.
Epidermis	Epidermis is the outermost layer of the skin. It forms the waterproof, protective wrap over the body's surface and is made up of stratified squamous epithelium with an underlying basement membrane. It contains no blood vessels, and is nourished by diffusion from the dermis. In plants, the outermost layer of cells covering the leaves and young parts of a plant is the epidermis.
Ground tissue system	Ground tissue system refers to a tissue of mostly parenchyma cells that makes up the bulk of a young plant and is continuous throughout its body.
Ground tissue	Tissue that constitutes most of the body of a plant is ground tissue.
Embryo	Embryo refers to a developing stage of a multicellular organism. In humans, the stage in the development of offspring from the first division of the zygote until body structures begin to appear, about the ninth week of gestation.

Cell division	Cell division is the process by which a cell (called the parent cell) divides into two cells (called daughter cells). Cell division is usually a small segment of a larger cell cycle. In meiosis, however, a cell is permanently transformed and cannot divide again.
Meristem	Meristem is a type of embryonic tissue in plants consisting of unspecialized, youthful cells called meristematic cells and found in areas of the plant where growth is or will take place - the roots and shoots.
Vascular cambium	Vascular cambium refers to during secondary growth of a plant, the cylinder of meristematic cells surrounding the xylem and pith that produces secondary xylem and phloem.
Cambium	A tissue in higher plants that produces new xylem and phloem is cambium.
Cork cambium	Cork cambium is one of the plant's meristems - the series of tissues consisting of embryonic (incompletely differentiated) cells from which the plant grows.
Cork	Cork is a tissue found in some plants, which consists of tightly packed dead cells. It allows improved insulation and prevents loss of water or nutrients on the inner bark of woody plants.
Root apical meristem	Undifferentiated, embryonic tissue located at the apex of the stem is root apical meristem.
Apical meristem	In vascular plants, the growing point at the tip of the root or stem is called the apical meristem.
Daughter cell	A cell formed by cell division of a parent cell is a daughter cell.
Root cap	At the tip of every growing root is a conical covering of tissue called the root cap. It usually is not visible to the naked eye.
Cortex	In anatomy and zoology the cortex is the outermost or superficial layer of an organ or the outer portion of the stem or root of a plant.
Pericycle	Found in the stele of plants, the pericycle is a cylinder of cells that lies just inside the endodermis. It conducts water and nutrients inward to the vascular tissue. In dicots, it is also has the capacity to produce lateral roots.
Shoot apical meristem	Undifferentiated tissue, located within the shoot tip of a plant, generally appearing as a shiny dome-like structure distal to the youngest leaf primordium and measuring less than 0.1 mm in length when excised is a shoot apical meristem.
Vascular bundle	A vascular bundle is a part of the transport system in vascular plants. The transport itself happens in vascular tissue, which exists in two forms: xylem and phloem. Both these tissues are present in a vascular bundle, which in addition will include supporting and protective tissues.
Pith	Pith refers to parenchyma tissue in the center of some stems and roots.
Annual ring	An annual ring can be seen in a horizontal cross section cut through the trunk of a tree. Visible rings result from the change in growth speed through the seasons of the year, thus one ring usually marks the passage of one year in the life of the tree. The rings are more visible in temperate zones, where the seasons differ more markedly.
Secondary growth	An increase in a plant's girth, involving cell division in the vascular cambium and cork cambium is secondary growth.
Secondary phloem	Phloem produced from the cells that arise toward the outside of the vascular cambium during secondary growth in a vascular plant is called secondary phloem.
Photosynthesis	Photosynthesis is a biochemical process in which plants, algae, and some bacteria harness the energy of light to produce food. Ultimately, nearly all living things depend on energy

produced from photosynthesis for their nourishment, making it vital to life on Earth.

Spongy mesophyll	Layer of tissue above the lower epidermis in a plant leaf containing loosely packed cells increasing the amount of surface area for gas exchange is referred to as spongy mesophyll.
Mesophyll	Mesophyll refers to the green tissue in the interior of a leaf; a leaf's ground tissue system, the main site of photosynthesis.
Cuticle	Cuticle in animals, a tough, nonliving outer layer of the skin. In plants, a waxy coating on the surface of stems and leaves that helps retain water.

Go to **Cram101.com** for the Practice Tests for this Chapter.

Xylem sap	The solution of water and inorganic nutrients (although it can contain a number of organic chemicals as well) conveyed in xylem tissue from a plant's roots to its shoots is referred to as xylem sap.
Xylem	The nonliving portion of a plant's vascular system that provides support and conveys water and inorganic nutrients from the roots to the rest of the plant. Xylem is made up of vessel elements and/or tracheids, water-conducting cells.
Gymnosperm	Gymnosperm refers to a naked-seed plant. Its seed is said to be naked because it is not enclosed in a fruit.
Solute	Substance that is dissolved in a solvent, forming a solution is referred to as a solute.
Osmosis	Osmosis is the diffusion of a solvent through a semipermeable membrane from a region of low solute concentration to a region of high solute concentration.
Cell wall	Cell wall refers to a protective layer external to the plasma membrane in plant cells, bacteria, fungi, and some protists; protects the cell and helps maintain its shape.
Cell	The cell is the structural and functional unit of all living organisms, and is sometimes called the "building block of life."
Ion	Ion refers to an atom or molecule that has gained or lost one or more electrons, thus acquiring an electrical charge.
Gradient	Gradient refers to a difference in concentration, pressure, or electrical charge between two regions.
Sodium-potassium pump	Carrier protein in the plasma membrane that moves sodium ions out of and potassium into cells is called sodium-potassium pump.
Proton	Positive subatomic particle, located in the nucleus and having a weight of approximately one atomic mass unit is referred to as a proton.
Bulk flow	The movement of many molecules of a gas or fluid in unison from an area of higher pressure to an area of lower pressure is referred to as bulk flow.
Plasma	In physics and chemistry, a plasma is an ionized gas, and is usually considered to be a distinct phase of matter. "Ionized" in this case means that at least one electron has been dissociated from a significant fraction of the molecules.
Root hair	Root hair refers to an outgrowth of an epidermal cell on a root, which increases the root's absorptive surface area.
Root	In vascular plants, the root is that organ of a plant body that typically lies below the surface of the soil. However, this is not always the case, since a root can also be aerial (that is, growing above the ground) or aerating (that is, growing up above the ground or especially above water).
Cortex	In anatomy and zoology the cortex is the outermost or superficial layer of an organ or the outer portion of the stem or root of a plant.
Cytoplasm	Cytoplasm refers to everything inside a cell between the plasma membrane and the nucleus; consists of a semifluid medium and organelles.
Selectively permeable	The property of a membrane that allows certain molecules to pass through it but interferes with the passage of others is referred to as selectively permeable.
Endodermis	Endodermis is the bottom layer of skin. In plants, it is a thin layer of parenchyma found in roots, just outside the vascular cylinder. It regulates the flow of water.
Pericycle	Found in the stele of plants, the pericycle is a cylinder of cells that lies just inside the

Go to **Cram101.com** for the Practice Tests for this Chapter.

endodermis. It conducts water and nutrients inward to the vascular tissue. In dicots, it is also has the capacity to produce lateral roots.

Bark	Bark is the outermost layer of stems and roots of woody plants such as trees. It overlays the wood and consists of three layers, the cork, the phloem, and the vascular cambium.
Root pressure	Root pressure is one of the phenomena used by vascular plants to move water into the leaves. The water in the soil tends to be poorer in solutes than the water in the plant's cells, due to the plant's active absorption of dissolved nutrients. The resulting solute potential gradient causes water to flow into the roots. Root pressure is capable, under ideal atmospheric conditions, of pushing water one or two feet above the ground.
Leaf	In botany, a leaf is an above-ground plant organ specialized for photosynthesis. For this purpose, a leaf is typically flat (laminar) and thin, to expose the chloroplast containing cells (chlorenchyma tissue) to light over a broad area, and to allow light to penetrate fully into the tissues.
Surface tension	A measure of how difficult it is to stretch or break the surface of a liquid is referred to as surface tension.
Mesophyll	Mesophyll refers to the green tissue in the interior of a leaf; a leaf's ground tissue system, the main site of photosynthesis.
Stem	Stem refers to that part of a plant's shoot system that supports the leaves and reproductive structures.
Species	Group of similarly constructed organisms capable of interbreeding and producing fertile offspring is a species.
Transpiration	Transpiration refers to plant's loss of water to the atmosphere, mainly through evaporation at leaf stomata.
Shoot	In botany, the shoot is one of two primary sections of a plant; the other is the root. The shoot refers to what is generally the upper portion of a plant, and consists of stems, leaves, flowers, and fruits. It is derived from the embryonic epicotyl, the portion of the embryo above the point of attachment to the seed leaves (cotyledons).
Epidermis	Epidermis is the outermost layer of the skin. It forms the waterproof, protective wrap over the body's surface and is made up of stratified squamous epithelium with an underlying basement membrane. It contains no blood vessels, and is nourished by diffusion from the dermis. In plants, the outermost layer of cells covering the leaves and young parts of a plant is the epidermis.
Cuticle	Cuticle in animals, a tough, nonliving outer layer of the skin. In plants, a waxy coating on the surface of stems and leaves that helps retain water.
Guard cell	A specialized epidermal cell in plants that regulates the size of a stoma, allowing gas exchange between the surrounding air and the photosynthetic cells in the leaf is called guard cell.
Photosynthesis	Photosynthesis is a biochemical process in which plants, algae, and some bacteria harness the energy of light to produce food. Ultimately, nearly all living things depend on energy produced from photosynthesis for their nourishment, making it vital to life on Earth.
Acid	An acid is a water-soluble, sour-tasting chemical compound that when dissolved in water, gives a solution with a pH of less than 7.
Stoma	A stoma is a tiny opening or pore, found mostly on the undersurface of a plant leaf, and used for gas exchange.
Pigment	Pigment is any material resulting in color in plant or animal cells which is the result of

Go to **Cram101.com** for the Practice Tests for this Chapter.

229

selective absorption.

Phloem	In vascular plants, phloem is the living tissue that carries organic nutrients, particularly sucrose, to all parts of the plant where needed. In trees, the phloem is underneath and difficult to distinguish from bark,
Respiration	Respiration is the process by which an organism obtains energy by reacting oxygen with glucose to give water, carbon dioxide and ATP (energy). Respiration takes place on a cellular level in the mitochondria of the cells and provide the cells with energy.
Translocation	A chromosomal mutation in which a portion of one chromosome breaks off and becomes attached to another chromosome is referred to as translocation.
Organ	Organ refers to a structure consisting of several tissues adapted as a group to perform specific functions.
Sieve tube	In phloem, a single strand of sievetube elements that transports sugar solutions is called sieve tube.
Protein	A protein is a complex, high-molecular-weight organic compound that consists of amino acids joined by peptide bonds. They are essential to the structure and function of all living cells and viruses. Many are enzymes or subunits of enzymes.
Tissue	Group of similar cells which perform a common function is called tissue.
Active transport	Active transport is the mediated transport of biochemicals, and other atomic/molecular substances, across membranes. In this form of transport, molecules move against either an electrical or concentration gradient.
Plasma membrane	Membrane surrounding the cytoplasm that consists of a phospholipid bilayer with embedded proteins is referred to as plasma membrane.

Leaf	In botany, a leaf is an above-ground plant organ specialized for photosynthesis. For this purpose, a leaf is typically flat (laminar) and thin, to expose the chloroplast containing cells (chlorenchyma tissue) to light over a broad area, and to allow light to penetrate fully into the tissues.
Adaptation	A biological adaptation is an anatomical structure, physiological process or behavioral trait of an organism that has evolved over a period of time by the process of natural selection such that it increases the expected long-term reproductive success of the organism.
Photosynthesis	Photosynthesis is a biochemical process in which plants, algae, and some bacteria harness the energy of light to produce food. Ultimately, nearly all living things depend on energy produced from photosynthesis for their nourishment, making it vital to life on Earth.
Light reactions	The first of two stages in photosynthesis, the steps in which solar energy is absorbed and converted to chemical energy in the form of ATP and NADPH. The light reactions power the sugar-producing Calvin cycle but produce no sugar themselves.
Light reaction	The first of two stages in photosynthesis, the steps in which solar energy is absorbed and converted to chemical energy in the form of ATP and NADPH. The light reaction powers the sugar-producing Calvin cycle but produces no sugar itself.
Bacteria	The domain that contains procaryotic cells with primarily diacyl glycerol diesters in their membranes and with bacterial rRNA. Bacteria also is a general term for organisms that are composed of procaryotic cells and are not multicellular.
Microorganism	A microorganism is an organism that is so small that it is microscopic (invisible to the naked eye). They are often illustrated using single-celled, or unicellular organisms; however, some unicellular protists are visible to the naked eye, and some multicellular species are microscopic.
Nucleic acid	A nucleic acid is a complex, high-molecular-weight biochemical macromolecule composed of nucleotide chains that convey genetic information. The most common are deoxyribonucleic acid (DNA) and ribonucleic acid (RNA). They are found in all living cells and viruses.
Acid	An acid is a water-soluble, sour-tasting chemical compound that when dissolved in water, gives a solution with a pH of less than 7.
Carnivore	An animal that eats a diet consisting solely of meat is referred to as a carnivore.
Sessile	In botany, sessile means without a stalk, as in flowers or leaves that grow directly from the stem. In zoology, sessile organisms are those which are not able to move about. They are usually permanently attached to a solid substrate of some kind, such as a rock, or the hull of a ship in the case of barnacles.
Root	In vascular plants, the root is that organ of a plant body that typically lies below the surface of the soil. However, this is not always the case, since a root can also be aerial (that is, growing above the ground) or aerating (that is, growing up above the ground or especially above water).
Element	A chemical element, often called simply element, is a chemical substance that cannot be divided or changed into other chemical substances by any ordinary chemical technique. An element is a class of substances that contain the same number of protons in all its atoms.
Chlorophyll	Chlorophyll is a green photosynthetic pigment found in plants, algae, and cyanobacteria. In plant photosynthesis incoming light is absorbed by chlorophyll and other accessory pigments in the antenna complexes of photosystem I and photosystem II.
Cytoskeleton	Cytoskeleton refers to a meshwork of fine fibers in the cytoplasm of a eukaryotic cell; includes microfilaments, intermediate filaments, and microtubules.

Nutrition	Nutrition refers to collectively, the processes involved in taking in, assimilating, and utilizing nutrients.
Ion	Ion refers to an atom or molecule that has gained or lost one or more electrons, thus acquiring an electrical charge.
Climate	Weather condition of an area including especially prevailing temperature and average daily/yearly rainfall over a long period of time is called climate.
Decomposition	Decomposition refers to the reduction of the body of a formerly living organism into simpler forms of matter.
Macronutrient	A chemical substance that an organism must obtain in relatively large amounts is referred to as macronutrient.
Nitrogen fixation	Nitrogen fixation is the process by which nitrogen is taken from its relatively inert molecular form (N_2) in the atmosphere and converted into nitrogen compounds useful for other chemical processes.
Nitrogen	Nitrogen is a chemical element in the periodic table which has the symbol N and atomic number 7. Commonly a colorless, odorless, tasteless and mostly inert diatomic non-metal gas, nitrogen constitutes 78.08% percent of Earth's atmosphere and is a constituent of all living tissues. Nitrogen forms many important compounds such as amino acids, ammonia, nitric acid, and cyanides.
Fixation	Fixation in population genetics occurs when the frequency of a gene reaches 1.Fixation in biochemistry, histology, cell biology and pathology refers to the technique of preserving a specimen for microscopic study, making it intact and stable, but dead.
Cyanobacteria	Cyanobacteria are a phylum of bacteria that obtain their energy through photosynthesis. They are often referred to as blue-green algae, even though it is now known that they are not directly related to any of the other algal groups, which are all eukaryotes.
Genus	In biology, a genus is a taxonomic grouping. That is, in the classification of living organisms, a genus is considered to be distinct from other such genera. A genus has one or more species: if it has more than one species these are likely to be morphologically more similar than species belonging to different genera.
Species	Group of similarly constructed organisms capable of interbreeding and producing fertile offspring is a species.
Legume	Legume refers to a member of a family of plants characterized by root swellings in which nitrogen-fixing bacteria are housed; includes soybeans, lupines, alfalfa, and clover.
Crop	An organ, found in both earthworms and birds, in which ingested food is temporarily stored before being passed to the gizzard, where it is pulverized is the crop.
Reactant	A reactant is any substance initially present in a chemical reaction. These reactants react with each other to form the products of a chemical reaction. In a chemical equation, the reactants are the elements or compounds on the left hand side of the reaction equation.
Root nodule	A swelling on a plant root consisting of plant cells that contains nitrogen-fixing bacteria is a root nodule.
Nodule	A swelling on the root of a legume or other plant that consists of cortex cells inhabited by nitrogen-fixing bacteria is a nodule.
Symbiosis	Symbiosis is an interaction between two organisms living together in more or less intimate association or even the merging of two dissimilar organisms. The term host is usually used for the larger (macro) of the two members of a symbiosis.

Go to **Cram101.com** for the Practice Tests for this Chapter.

Transcription	Transcription is the process through which a DNA sequence is enzymatically copied by an RNA polymerase to produce a complementary RNA. Or, in other words, the transfer of genetic information from DNA into RNA.
Protein	A protein is a complex, high-molecular-weight organic compound that consists of amino acids joined by peptide bonds. They are essential to the structure and function of all living cells and viruses. Many are enzymes or subunits of enzymes.
Hemoglobin	Hemoglobin is the iron-containing oxygen-transport metalloprotein in the red cells of the blood in mammals and other animals. Hemoglobin transports oxygen from the lungs to the rest of the body, such as to the muscles, where it releases the oxygen load.
Respiration	Respiration is the process by which an organism obtains energy by reacting oxygen with glucose to give water, carbon dioxide and ATP (energy). Respiration takes place on a cellular level in the mitochondria of the cells and provide the cells with energy.
Bacterium	Most bacterium are microscopic and unicellular, with a relatively simple cell structure lacking a cell nucleus, and organelles such as mitochondria and chloroplasts. They are the most abundant of all organisms. They are ubiquitous in soil, water, and as symbionts of other organisms.
Recombinant DNA	Recombinant DNA is an artificial DNA sequence resulting from the combining of two other DNA sequences in a plasmid.
Angiosperm	Flowering plant that produces seeds within an ovary that develops into a fruit is referred to as an angiosperm.
Ammonia	Ammonia is a compound of nitrogen and hydrogen with the formula NH_3. At standard temperature and pressure ammonia is a gas. It is toxic and corrosive to some materials, and has a characteristic pungent odor.
Metabolism	Metabolism is the biochemical modification of chemical compounds in living organisms and cells. This includes the biosynthesis of complex organic molecules (anabolism) and their breakdown (catabolism).
Coenzyme	Nonprotein organic molecule that aids the action of the enzyme to which it is loosely bound is referred to as coenzyme.
Evolution	In biology, evolution is the process by which novel traits arise in populations and are passed on from generation to generation. Its action over large stretches of time explains the origin of new species and ultimately the vast diversity of the biological world.
Enzyme	An enzyme is a protein that catalyzes, or speeds up, a chemical reaction. They are essential to sustain life because most chemical reactions in biological cells would occur too slowly, or would lead to different products, without them.

Go to **Cram101.com** for the Practice Tests for this Chapter.

Stem	Stem refers to that part of a plant's shoot system that supports the leaves and reproductive structures.
Fruit	A fruit is the ripened ovary—together with seeds—of a flowering plant. In many species, the fruit incorporates the ripened ovary and surrounding tissues.
Hormone	A hormone is a chemical messenger from one cell to another. All multicellular organisms produce hormones. The best known hormones are those produced by endocrine glands of vertebrate animals, but hormones are produced by nearly every organ system and tissue type in a human or animal body. Hormone molecules are secreted directly into the bloodstream, they move by circulation or diffusion to their target cells, which may be nearby cells in the same tissue or cells of a distant organ of the body.
Plant hormone	Chemical signal that is produced by various plant tissues and coordinates the activities of plant cells is referred to as plant hormone.
Genome	The genome of an organism is the whole hereditary information of an organism that is encoded in the DNA (or, for some viruses, RNA). This includes both the genes and the non-coding sequences. The genome of an organism is a complete DNA sequence of one set of chromosomes.
Receptor	A receptor is a protein on the cell membrane or within the cytoplasm or cell nucleus that binds to a specific molecule (a ligand), such as a neurotransmitter, hormone, or other substance, and initiates the cellular response to the ligand. Receptor, in immunology, the region of an antibody which shows recognition of an antigen.
Cell division	Cell division is the process by which a cell (called the parent cell) divides into two cells (called daughter cells). Cell division is usually a small segment of a larger cell cycle. In meiosis, however, a cell is permanently transformed and cannot divide again.
Cell	The cell is the structural and functional unit of all living organisms, and is sometimes called the "building block of life."
Embryo	Embryo refers to a developing stage of a multicellular organism. In humans, the stage in the development of offspring from the first division of the zygote until body structures begin to appear, about the ninth week of gestation.
Species	Group of similarly constructed organisms capable of interbreeding and producing fertile offspring is a species.
Shoot	In botany, the shoot is one of two primary sections of a plant; the other is the root. The shoot refers to what is generally the upper portion of a plant, and consists of stems, leaves, flowers, and fruits. It is derived from the embryonic epicotyl, the portion of the embryo above the point of attachment to the seed leaves (cotyledons).
Fertilization	Fertilization is fusion of gametes to form a new organism. In animals, the process involves a sperm fusing with an ovum, which eventually leads to the development of an embryo.
Ethylene	Ethylene functions as a hormone in plants. It stimulates Ethylene (or IUPAC name ethene) is the simplest alkene hydrocarbon, consisting of four hydrogen atoms and two carbon atoms connected by a double bond.
Dormancy	Dormancy is a period of arrested plant growth. It is a survival strategy exhibited by many plant species, which enables them to survive in climates where part of the year is unsuitable for growth.
Microorganism	A microorganism is an organism that is so small that it is microscopic (invisible to the naked eye). They are often illustrated using single-celled, or unicellular organisms; however, some unicellular protists are visible to the naked eye, and some multicellular species are microscopic.

Go to **Cram101.com** for the Practice Tests for this Chapter.

Germination	The resumption of growth and development by a spore or seed is called germination. It is the process in botany where growth emerges from a resting stage.
Root	In vascular plants, the root is that organ of a plant body that typically lies below the surface of the soil. However, this is not always the case, since a root can also be aerial (that is, growing above the ground) or aerating (that is, growing up above the ground or especially above water).
Endosperm	Endosperm is a triploid tissue (containing three sets of chromosomes) found in the seeds of flowering plants. It provides nutrition to the developing embryo. It is mostly composed of starch, though it can also contain oils and protein.
Starch	Biochemically, starch is a combination of two polymeric carbohydrates (polysaccharides) called amylose and amylopectin.
Polymer	Polymer is a generic term used to describe a very long molecule consisting of structural units and repeating units connected by covalent chemical bonds.
Digestion	Digestion refers to the mechanical and chemical breakdown of food into molecules small enough for the body to absorb; the second main stage of food processing, following ingestion.
Glucose	Glucose, a simple monosaccharide sugar, is one of the most important carbohydrates and is used as a source of energy in animals and plants. Glucose is one of the main products of photosynthesis and starts respiration.
Enzyme	An enzyme is a protein that catalyzes, or speeds up, a chemical reaction. They are essential to sustain life because most chemical reactions in biological cells would occur too slowly, or would lead to different products, without them.
Gibberellin	One of a family of plant hormones that trigger the germination of seeds and interact with auxins in regulating growth and fruit development is referred to as gibberellin.
Biennial	A plant that completes its life cycle in two years is a biennial.
Flower	A flower is the reproductive structure of a flowering plant. The flower structure contains the plant's reproductive organs, and its function is to produce seeds through sexual reproduction.
Shoot system	Shoot system refers to aboveground portion of a plant consisting of the stem, leaves, and flowers.
Auxin	Plant hormone regulating growth, particularly cell elongation is referred to as an auxin. They have been demonstrated to be the basic coordinative signal of plant development. Their transport throughout plants is complex, and often they also control action of other plant hormones.
Leaf	In botany, a leaf is an above-ground plant organ specialized for photosynthesis. For this purpose, a leaf is typically flat (laminar) and thin, to expose the chloroplast containing cells (chlorenchyma tissue) to light over a broad area, and to allow light to penetrate fully into the tissues.
Carrier protein	Protein molecule that combines with a substance and transports it through the plasma membrane is called a carrier protein.
Protein	A protein is a complex, high-molecular-weight organic compound that consists of amino acids joined by peptide bonds. They are essential to the structure and function of all living cells and viruses. Many are enzymes or subunits of enzymes.
Petiole	Part of a plant leaf that connects the blade to the stem is called petiole.
Abscission	Abscission is the shedding of a body part. It most commonly refers to the process by which a

241

plant intentionally drops one or more of its parts, such as a leaf, fruit, flower or seed, though the term is also used to describe the shedding of a claw by an animal.

Apical dominance	The phenomenon whereby a growing shoot tip inhibits the sprouting of lateral buds is referred to as apical dominance.
Dominance	Dominance is the state of having high social status relative to other individuals, who react submissively to dominant individuals.
Acid	An acid is a water-soluble, sour-tasting chemical compound that when dissolved in water, gives a solution with a pH of less than 7.
Cellulose	A large polysaccharide composed of many glucose monomers linked into cable-like fibrils that provide structural support in plant cell walls is referred to as cellulose.
Cell wall	Cell wall refers to a protective layer external to the plasma membrane in plant cells, bacteria, fungi, and some protists; protects the cell and helps maintain its shape.
Vacuole	A vacuole is a large membrane-bound compartment within some eukaryotic cells where they serve a variety of different functions: capturing food materials or unwanted structural debris surrounding the cell, sequestering materials that might be toxic to the cell, maintaining fluid balance (called turgor) within the cell.
Cytoplasm	Cytoplasm refers to everything inside a cell between the plasma membrane and the nucleus; consists of a semifluid medium and organelles.
Polysaccharide	Polymer made from sugar monomers is a polysaccharide. They are relatively complex carbohydrates.
Gene	Gene refers to a discrete unit of hereditary information consisting of a specific nucleotide sequence in DNA . Most of the genes of a eukaryote are located in its chromosomal DNA; a few are carried by the DNA of mitochondria and chloroplasts.
Tissue	Group of similar cells which perform a common function is called tissue.
Pith	Pith refers to parenchyma tissue in the center of some stems and roots.
Vascular tissue	Plant tissue consisting of cells joined into tubes that transport water and nutrients throughout the plant body is referred to as vascular tissue.
Vascular	In botany vascular refers to tissues that contain vessels for transporting liquids. In anatomy and physiology, vascular means related to blood vessels, which are part of the Circulatory system.
Organ	Organ refers to a structure consisting of several tissues adapted as a group to perform specific functions.
Cytokinin	A cytokinin is a class of plant growth substances active in promoting cell division, and it is also involved in cell growth, differentiation, and other physiological processes.
Adenine	Adenine is one of the two purine nucleobases used in forming nucleotides of the nucleic acids DNA and RNA. In DNA, adenine (A) binds to thymine (T) via two hydrogen bonds to assist in stabilizing the nucleic acid structures. In RNA, adenine binds to uracil (U).
Senescence	Senescence is the combination of processes of deterioration which follow the period of development of an organism.
Petal	A petal is one member or part of the corolla of a flower. It is the inner part of the perianth that comprises the sterile parts of a flower and consists of inner and outer tepals.
Salt	Salt is a term used for ionic compounds composed of positively charged cations and negatively charged anions, so that the product is neutral and without a net charge.

Go to **Cram101.com** for the Practice Tests for this Chapter.

Abscisic acid	Abscisic acid is defined as a plant growth regulator that acts mainly to inhibit growth, promote dormancy, and help the plant tolerate stressful conditions.
Allele	An allele is any one of a number of viable DNA codings of the same gene (sometimes the term refers to a non-gene sequence) occupying a given locus (position) on a chromosome.
Mutation	Mutation refers to a change in the nucleotide sequence of DNA; the ultimate source of genetic diversity.
Wavelength	The distance between crests of adjacent waves, such as those of the electromagnetic spectrum is wavelength.
Photoreceptor	A photoreceptor is a specialized type of neuron that is capable of phototransduction. More specifically, the photoreceptor sends signals to other neurons by a change in its membrane potential when it absorbs photons.
Radicle	The radicle is the first part of a seedling to emerge from the seed during germination. The radicle is the embryonic root of the plant, and grows downward in the soil.
Chlorophyll	Chlorophyll is a green photosynthetic pigment found in plants, algae, and cyanobacteria. In plant photosynthesis incoming light is absorbed by chlorophyll and other accessory pigments in the antenna complexes of photosystem I and photosystem II.
Phytochrome	Phytochrome refers to photoreversible plant pigment that plants use to detect light and is involved in photoperiodism and other responses of plants such as etiolation.
Transduction	In physiology, transduction is transportation of a stimuli to the nervous system. In genetics, transduction is the transfer of viral, bacterial, or both bacterial and viral DNA from one cell to another via bacteriophage.
Nucleus	In cell biology, the nucleus is found in all eukaryotic cells that contains most of the cell's genetic material. The nucleus has two primary functions: to control chemical reactions within the cytoplasm and to store information needed for cellular division.
Signal transduction pathway	In cell biology, a series of molecular changes that converts a signal on a target cell's surface into a specific response inside the cell is referred to as signal transduction pathway.
Signal transduction	Signal transduction is any process by which a cell converts one kind of signal or stimulus into another. Processes referred to as signal transduction often involve a sequence of biochemical reactions inside the cell, which are carried out by enzymes and linked through second messengers.

Reproduction	Biological reproduction is the biological process by which new individual organisms are produced. Reproduction is a fundamental feature of all known life; each individual organism exists as the result of reproduction by an antecedent.
Asexual reproduction	Asexual reproduction refers to reproduction that does not involve the fusion of haploid sex cells. The parent body may divide and new parts regenerate, or a new, smaller individual may form as an attachment to the parent, to drop off when complete.
Population	Group of organisms of the same species occupying a certain area and sharing a common gene pool is referred to as population.
Genetic diversity	Genetic diversity is a characteristic of ecosystems and gene pools that describes an attribute which is commonly held to be advantageous for survival -- that there are many different versions of otherwise similar organisms.
Complete flower	A flower that has all four floral parts(stamen, a pistil, petals, and sepals) is referred to as a complete flower.
Flower	A flower is the reproductive structure of a flowering plant. The flower structure contains the plant's reproductive organs, and its function is to produce seeds through sexual reproduction.
Pistil	Structure in a flower that consists of a stigma, a style, and an ovule-containing ovary is called pistil.
Stamen	The stamen is the male organ of a flower. Each stamen generally has a stalk called the filament, and, on top of the filament, an anther.
Petal	A petal is one member or part of the corolla of a flower. It is the inner part of the perianth that comprises the sterile parts of a flower and consists of inner and outer tepals.
Sporophyte	A sporophyte is the diploid structure or phase of life of a sexually reproducing plant. Each living cell of the sporophyte contains two complete sets of chromosomes. The sporophyte is the dominant life form in ferns, gymnosperms, and angiosperms.
Spore	Spore refers to a differentiated, specialized form that can be used for dissemination, for survival of adverse conditions because of its heat and dessication resistance, and/or for reproduction. They are usually unicellular and may develop into vegetative organisms or gametes. A spore may be produced asexually or sexually and are of many types.
Gamete	A gamete is a specialized germ cell that unites with another gamete during fertilization in organisms that reproduce sexually. They are haploid cells; that is, they contain one complete set of chromosomes. When they unite they form a zygote—a cell having two complete sets of chromosomes and therefore diploid.
Embryo	Embryo refers to a developing stage of a multicellular organism. In humans, the stage in the development of offspring from the first division of the zygote until body structures begin to appear, about the ninth week of gestation.
Pollen	The male gametophyte in gymnosperms and angiosperms is referred to as pollen.
Ovule	An ovule is a structure found in seed plants that develops into a seed after fertilization. It consists of an embryo sac containing the female gamete, an ovum or egg cell, surrounded by nutritive tissue, the nucellus. Outside this there are one or two coverings that provide protection, developing into the testa, or seed coat, following fertilization.
Megaspore	Macrosporangium or megasporangium, is a type of spore sac produced by ferns, which contains megaspore (the fist cells of the female gametophyte generation).
Cell wall	Cell wall refers to a protective layer external to the plasma membrane in plant cells, bacteria, fungi, and some protists; protects the cell and helps maintain its shape.

Go to **Cram101.com** for the Practice Tests for this Chapter.

Cell	The cell is the structural and functional unit of all living organisms, and is sometimes called the "building block of life."
Megagametophyte	Megagametophyte in seed plants, is the gametophyte that produces an egg.
Egg	An egg is the zygote, resulting from fertilization of the ovum. It nourishes and protects the embryo.
Haploid	Haploid cells bear one copy of each chromosome.
Angiosperm	Flowering plant that produces seeds within an ovary that develops into a fruit is referred to as an angiosperm.
Gymnosperm	Gymnosperm refers to a naked-seed plant. Its seed is said to be naked because it is not enclosed in a fruit.
Pollination	In seed plants, the delivery of pollen to the vicinity of the egg-producing megagametophyte is pollination.
Anther	The stamen is the male organ of a flower. Each stamen generally has a stalk called the filament, and, on top of the filament, an anther. The anther is usually composed of four pollen sacs, which are called microsporangia.
Gene	Gene refers to a discrete unit of hereditary information consisting of a specific nucleotide sequence in DNA . Most of the genes of a eukaryote are located in its chromosomal DNA; a few are carried by the DNA of mitochondria and chloroplasts.
Pollen grain	Pollen grain in seed plants, the sperm-producing microgametophyte.
Stigma	The sticky tip of a flower's carpel, which traps pollen grains is called stigma. In zoology, one of the external openings of the tracheae of insects, myriapods, and other arthropods; a spiracle. One of the apertures of the pulmonary sacs of arachnids. One of the apertures of the gill of an ascidian, and of Amphioxus.
Species	Group of similarly constructed organisms capable of interbreeding and producing fertile offspring is a species.
Pollen tube	A tube that grows from a pollen grain. Male reproductive cells move through the pollen tube into the ovule.
Tube cell	Tube cell refers to the outermost cell of a pollen grain; digests a tube through the tissues of the carpel, ultimately penetrating into the female gametophyte.
Generative cell	Generative cell in flowering plants, one of the haploid cells of a pollen grain; undergoes mitosis to form two sperm cells.
Sperm	Sperm refers to the male sex cell with three distinct parts at maturity: head, middle piece, and tail.
Zygote	Diploid cell formed by the union of sperm and egg is referred to as zygote.
Fertilization	Fertilization is fusion of gametes to form a new organism. In animals, the process involves a sperm fusing with an ovum, which eventually leads to the development of an embryo.
Seed coat	Protective covering for the seed, formed by the hardening of the ovule wall is a seed coat.
Cytoplasm	Cytoplasm refers to everything inside a cell between the plasma membrane and the nucleus; consists of a semifluid medium and organelles.
Hypocotyl	Hypocotyl is a botanical term for a part of a germinating seedling of a seed plant. As the plant embryo grows at germination, it sends out a shoot called a radicle that becomes the primary root and penetrates down into the soil.

Go to **Cram101.com** for the Practice Tests for this Chapter.

Fruit	A fruit is the ripened ovary—together with seeds—of a flowering plant. In many species, the fruit incorporates the ripened ovary and surrounding tissues.
Shoot system	Shoot system refers to aboveground portion of a plant consisting of the stem, leaves, and flowers.
Shoot	In botany, the shoot is one of two primary sections of a plant; the other is the root. The shoot refers to what is generally the upper portion of a plant, and consists of stems, leaves, flowers, and fruits. It is derived from the embryonic epicotyl, the portion of the embryo above the point of attachment to the seed leaves (cotyledons).
Apical meristem	In vascular plants, the growing point at the tip of the root or stem is called the apical meristem.
Meristem	Meristem is a type of embryonic tissue in plants consisting of unspecialized, youthful cells called meristematic cells and found in areas of the plant where growth is or will take place - the roots and shoots.
Determinate growth	Termination of growth after reaching a certain size, as in most animals is determinate growth.
Determinate	Determinate refers to a condition when the terminal bud ceases to grow, preventing elongation of the main axis and promoting auxiliary growth.
Short-day plant	Short-day plant refers to plant which flowers when day length is shorter than a critical length, i.e., cocklebur, poinsettia, and chrysanthemum.
Biological clock	Biological clock refers to an internal timekeeper that controls an organism's biological rhythms; marks time with or without environmental cues but often requires signals from the environment to remain tuned to an appropriate period.
Stimulus	Stimulus in a nervous system, a factor that triggers sensory transduction.
Leaf	In botany, a leaf is an above-ground plant organ specialized for photosynthesis. For this purpose, a leaf is typically flat (laminar) and thin, to expose the chloroplast containing cells (chlorenchyma tissue) to light over a broad area, and to allow light to penetrate fully into the tissues.
Hormone	A hormone is a chemical messenger from one cell to another. All multicellular organisms produce hormones. The best known hormones are those produced by endocrine glands of vertebrate animals, but hormones are produced by nearly every organ system and tissue type in a human or animal body. Hormone molecules are secreted directly into the bloodstream, they move by circulation or diffusion to their target cells, which may be nearby cells in the same tissue or cells of a distant organ of the body.
Florigen	Florigen is the term used for the hypothesized hormone-like molecules that control and/or trigger flowering in plants. Its precise identity and mechanism are not known; only its function.
Genetic recombination	Genetic recombination refers to the production, by crossing over and/or independent assortment of chromosomes during meiosis, of offspring with allele combinations different from those in the parents.
Recombination	Genetic recombination is the transmission-genetic process by which the combinations of alleles observed at different loci in two parental individuals become shuffled in offspring individuals.
Clone	A group of genetically identical cells or organisms derived by asexual reproduction from a single parent is called a clone.
Organ	Organ refers to a structure consisting of several tissues adapted as a group to perform

	specific functions.
Stem	Stem refers to that part of a plant's shoot system that supports the leaves and reproductive structures.
Meiosis	In biology, meiosis is the process that transforms one diploid cell into four haploid cells in eukaryotes in order to redistribute the diploid's cell's genome. Meiosis forms the basis of sexual reproduction and can only occur in eukaryotes.
Gametophyte	Gametophyte refers to the multicellular haploid form in the life cycle of organisms undergoing alternation of generations; mitotically produces haploid gametes that unite and grow into the sporophyte generation.
Cell division	Cell division is the process by which a cell (called the parent cell) divides into two cells (called daughter cells). Cell division is usually a small segment of a larger cell cycle. In meiosis, however, a cell is permanently transformed and cannot divide again.
Cambium	A tissue in higher plants that produces new xylem and phloem is cambium.
Root	In vascular plants, the root is that organ of a plant body that typically lies below the surface of the soil. However, this is not always the case, since a root can also be aerial (that is, growing above the ground) or aerating (that is, growing up above the ground or especially above water).
Genus	In biology, a genus is a taxonomic grouping. That is, in the classification of living organisms, a genus is considered to be distinct from other such genera. A genus has one or more species: if it has more than one species these are likely to be morphologically more similar than species belonging to different genera.
Totipotent	Cell that has the full genetic potential of the organism and has the potential to develop into a complete organism is referred to as totipotent.
Vascular tissue	Plant tissue consisting of cells joined into tubes that transport water and nutrients throughout the plant body is referred to as vascular tissue.
Vascular	In botany vascular refers to tissues that contain vessels for transporting liquids. In anatomy and physiology, vascular means related to blood vessels, which are part of the Circulatory system.
Tissue	Group of similar cells which perform a common function is called tissue.

Herbivore	A herbivore is an animal that is adapted to eat primarily plant matter
Pathogen	A pathogen or infectious agent is a biological agent that causes disease or illness to its host.The term is most often used for agents that disrupt the normal physiology of a multicellular animal or plant.
Epidermis	Epidermis is the outermost layer of the skin. It forms the waterproof, protective wrap over the body's surface and is made up of stratified squamous epithelium with an underlying basement membrane. It contains no blood vessels, and is nourished by diffusion from the dermis. In plants, the outermost layer of cells covering the leaves and young parts of a plant is the epidermis.
Cell	The cell is the structural and functional unit of all living organisms, and is sometimes called the "building block of life."
Lignin	Lignin is a chemical compound that is most commonly derived from wood and is an integral part of the cell walls of plants, especially in tracheids, xylem fibres and sclereids. It is the second most abundant organic compound on earth after cellulose. Lignin makes up about one-quarter to one-third of the dry mass of wood.
Bacteria	The domain that contains procaryotic cells with primarily diacyl glycerol diesters in their membranes and with bacterial rRNA. Bacteria also is a general term for organisms that are composed of procaryotic cells and are not multicellular.
Species	Group of similarly constructed organisms capable of interbreeding and producing fertile offspring is a species.
Tissue	Group of similar cells which perform a common function is called tissue.
Acid	An acid is a water-soluble, sour-tasting chemical compound that when dissolved in water, gives a solution with a pH of less than 7.
Gene	Gene refers to a discrete unit of hereditary information consisting of a specific nucleotide sequence in DNA . Most of the genes of a eukaryote are located in its chromosomal DNA; a few are carried by the DNA of mitochondria and chloroplasts.
Dominant allele	Dominant allele refers to an allele that exerts its phenotypic effect in the heterozygote.
Allele	An allele is any one of a number of viable DNA codings of the same gene (sometimes the term refers to a non-gene sequence) occupying a given locus (position) on a chromosome.
Microorganism	A microorganism is an organism that is so small that it is microscopic (invisible to the naked eye). They are often illustrated using single-celled, or unicellular organisms; however, some unicellular protists are visible to the naked eye, and some multicellular species are microscopic.
Predator	A predator is an animal or other organism that hunts and kills other organisms for food in an act called predation.
Natural selection	Natural selection is the process by which biological individuals that are endowed with favorable or deleterious traits end up reproducing more or less than other individuals that do not possess such traits.
Photosynthesis	Photosynthesis is a biochemical process in which plants, algae, and some bacteria harness the energy of light to produce food. Ultimately, nearly all living things depend on energy produced from photosynthesis for their nourishment, making it vital to life on Earth.
Shoot	In botany, the shoot is one of two primary sections of a plant; the other is the root. The shoot refers to what is generally the upper portion of a plant, and consists of stems, leaves, flowers, and fruits. It is derived from the embryonic epicotyl, the portion of the embryo above the point of attachment to the seed leaves (cotyledons).

Go to **Cram101.com** for the Practice Tests for this Chapter.

Pollen	The male gametophyte in gymnosperms and angiosperms is referred to as pollen.
Insect	An arthropod that usually has three body segments , three pairs of legs, and one or two pairs of wings is called an insect. They are the largest and (on land) most widely-distributed taxon within the phylum Arthropoda. They comprise the most diverse group of animals on the earth, with around 925,000 species described
Polypeptide	Polypeptide refers to polymer of many amino acids linked by peptide bonds.
Unsaturated fatty acid	Fatty acid molecule that has one or more double bonds between the atoms of its carbon chain is called unsaturated fatty acid.
Fatty acid	A fatty acid is a carboxylic acid (or organic acid), often with a long aliphatic tail (long chains), either saturated or unsaturated.
Inhibitor	An inhibitor is a type of effector (biology) that decreases or prevents the rate of a chemical reaction. They are often called negative catalysts.
Crop	An organ, found in both earthworms and birds, in which ingested food is temporarily stored before being passed to the gizzard, where it is pulverized is the crop.
Enzyme	An enzyme is a protein that catalyzes, or speeds up, a chemical reaction. They are essential to sustain life because most chemical reactions in biological cells would occur too slowly, or would lead to different products, without them.
Hydrophobic	Hydrophobic refers to being electrically neutral and nonpolar, and thus prefering other neutral and nonpolar solvents or molecular environments. Hydrophobic is often used interchangeably with "oily" or "lipophilic."
Vein	Vein in animals, is a vessel that returns blood to the heart. In plants, a vascular bundle in a leaf, composed of xylem and phloem.
Desert	A desert is a landscape form or region that receives little precipitation - less than 250 mm (10 in) per year. It is a biome characterized by organisms adapted to sparse rainfall and rapid evaporation.
Secretion	Secretion is the process of segregating, elaborating, and releasing chemicals from a cell, or a secreted chemical substance or amount of substance.
Adaptation	A biological adaptation is an anatomical structure, physiological process or behavioral trait of an organism that has evolved over a period of time by the process of natural selection such that it increases the expected long-term reproductive success of the organism.
Radiation	The emission of electromagnetic waves by all objects warmer than absolute zero is referred to as radiation.
Root	In vascular plants, the root is that organ of a plant body that typically lies below the surface of the soil. However, this is not always the case, since a root can also be aerial (that is, growing above the ground) or aerating (that is, growing up above the ground or especially above water).
Diffusion	Diffusion refers to the spontaneous movement of particles of any kind from where they are more concentrated to where they are less concentrated.
Respiration	Respiration is the process by which an organism obtains energy by reacting oxygen with glucose to give water, carbon dioxide and ATP (energy). Respiration takes place on a cellular level in the mitochondria of the cells and provide the cells with energy.
Leaf	In botany, a leaf is an above-ground plant organ specialized for photosynthesis. For this purpose, a leaf is typically flat (laminar) and thin, to expose the chloroplast containing cells (chlorenchyma tissue) to light over a broad area, and to allow light to penetrate fully

Go to **Cram101.com** for the Practice Tests for this Chapter.

	into the tissues.
Salt	Salt is a term used for ionic compounds composed of positively charged cations and negatively charged anions, so that the product is neutral and without a net charge.
Ion	Ion refers to an atom or molecule that has gained or lost one or more electrons, thus acquiring an electrical charge.
Amino acid	An amino acid is any molecule that contains both amino and carboxylic acid functional groups. They are the basic structural building units of proteins. They form short polymer chains called peptides or polypeptides which in turn form structures called proteins.
Habitat	Habitat refers to a place where an organism lives; an environmental situation in which an organism lives.

259

Desert	A desert is a landscape form or region that receives little precipitation - less than 250 mm (10 in) per year. It is a biome characterized by organisms adapted to sparse rainfall and rapid evaporation.
Circulatory system	The circulatory system or cardiovascular system is the organ system which circulates blood around the body of most animals.
Cell	The cell is the structural and functional unit of all living organisms, and is sometimes called the "building block of life."
Organ	Organ refers to a structure consisting of several tissues adapted as a group to perform specific functions.
Multicellular	Multicellular organisms are those organisms consisting of more than one cell, and having differentiated cells that perform specialized functions. Most life that can be seen with the naked eye is multicellular, as are all animals (i.e. members of the kingdom Animalia) and plants (i.e. members of the kingdom Plantae).
Sponge	An invertebrates that consist of a complex aggregation of cells, including collar cells, and has a skeleton of fibers and/or spicules is a sponge. They are primitive, sessile, mostly marine, waterdwelling filter feeders that pump water through their matrix to filter out particulates of food matter.
Evolution	In biology, evolution is the process by which novel traits arise in populations and are passed on from generation to generation. Its action over large stretches of time explains the origin of new species and ultimately the vast diversity of the biological world.
Salt	Salt is a term used for ionic compounds composed of positively charged cations and negatively charged anions, so that the product is neutral and without a net charge.
Homeostasis	Homeostasis is the property of an open system, especially living organisms, to regulate its internal environment to maintain a stable, constant condition, by means of multiple dynamic equilibrium adjustments, controlled by interrelated regulation mechanisms.
Tissue	Group of similar cells which perform a common function is called tissue.
Enzyme	An enzyme is a protein that catalyzes, or speeds up, a chemical reaction. They are essential to sustain life because most chemical reactions in biological cells would occur too slowly, or would lead to different products, without them.
Nervous system	The nervous system of an animal coordinates the activity of the muscles, monitors the organs, constructs and processes input from the senses, and initiates actions.
Epithelial	Functions of epithelial cells include secretion, absorption, protection, transcellular transport, sensation detection, and selective permeability.
Epithelium	Epithelium is a tissue composed of a layer of cells. Epithelium can be found lining internal (e.g. endothelium, which lines the inside of blood vessels) or external (e.g. skin) free surfaces of the body. Functions include secretion, absorption and protection.
Organelle	In cell biology, an organelle is one of several structures with specialized functions, suspended in the cytoplasm of a eukaryotic cell.
Skin	Skin is an organ of the integumentary system composed of a layer of tissues that protect underlying muscles and organs.
Extracellular matrix	Extracellular matrix is any material part of a tissue that is not part of any cell. Extracellular matrix is the defining feature of connective tissue.
Matrix	In biology, matrix (plural: matrices) is the material between animal or plant cells, the material (or tissue) in which more specialized structures are embedded, and a specific part

Go to **Cram101.com** for the Practice Tests for this Chapter.

	of the mitochondrion that is the site of oxidation of organic molecules.
Collagen	Collagen is the main protein of connective tissue in animals and the most abundant protein in mammals, making up about 1/4 of the total. It is one of the long, fibrous structural proteins whose functions are quite different from those of globular proteins such as enzymes.
Connective tissue	Connective tissue is any type of biological tissue with an extensive extracellular matrix and often serves to support, bind together, and protect organs.
Protein	A protein is a complex, high-molecular-weight organic compound that consists of amino acids joined by peptide bonds. They are essential to the structure and function of all living cells and viruses. Many are enzymes or subunits of enzymes.
Cartilage	Cartilage is a type of dense connective tissue. Cartilage is composed of cells called chondrocytes which are dispersed in a firm gel-like ground substance, called the matrix. Cartilage is avascular (contains no blood vessels) and nutrients are diffused through the matrix.
Skeletal system	Skeletal systems are commonly divided into three types - external (an exoskeleton), internal (an endoskeleton), and fluid based (a hydrostatic skeleton), though hydrostatic skeletal systems may be classified separately from the other two since they lack hardened support structures.
Vertebrate	Vertebrate is a subphylum of chordates, specifically, those with backbones or spinal columns. They started to evolve about 530 million years ago during the Cambrian explosion, which is part of the Cambrian period.
Blood	Blood is a circulating tissue composed of fluid plasma and cells. The main function of blood is to supply nutrients (oxygen, glucose) and constitutional elements to tissues and to remove waste products.
Muscle	Muscle is a contractile form of tissue. It is one of the four major tissue types, the other three being epithelium, connective tissue and nervous tissue. Muscle contraction is used to move parts of the body, as well as to move substances within the body.
Stomach	The stomach is an organ in the alimentary canal used to digest food. It's primary function is not the absorption of nutrients from digested food; rather, the main job of the stomach is to break down large food molecules into smaller ones, so that they can be absorbed into the blood more easily.
Negative feedback	Negative feedback refers to a control mechanism in which a chemical reaction, metabolic pathway, or hormonesecreting gland is inhibited by the products of the reaction, pathway, or gland.
Positive feedback	Mechanism of homeostatic response in which the output intensifies and increases the likelihood of response, instead of countering it and canceling it is called positive feedback.
Algae	The algae consist of several different groups of living organisms that capture light energy through photosynthesis, converting inorganic substances into simple sugars with the captured energy.
Species	Group of similarly constructed organisms capable of interbreeding and producing fertile offspring is a species.
Homeotherm	Homeotherm refers to an organism that maintains a stable internal body temperature regardless of external influence. This temperature is often higher than the immediate environment
Domain	In biology, a domain is the top-level grouping of organisms in scientific classification.
Basal metabolic	Basal metabolic rate, is the rate of metabolism that occurs when an individual is at rest in

263

rate	a warm environment and is in the post absorptive state, and has not eaten for at least 12 hours.
Metabolic rate	Energy expended by the body per unit time is called metabolic rate.
Climate	Weather condition of an area including especially prevailing temperature and average daily/yearly rainfall over a long period of time is called climate.
Brain	The part of the central nervous system involved in regulating and controlling body activity and interpreting information from the senses transmitted through the nervous system is referred to as the brain.
Hypothalamus	Located below the thalamus, the hypothalamus links the nervous system to the endocrine system by synthesizing and secreting neurohormones often called releasing hormones because they function by stimulating the secretion of hormones from the anterior pituitary gland.
Fever	Fever (also known as pyrexia, or a febrile response, and archaically known as ague) is a medical symptom that describes an increase in internal body temperature to levels that are above normal (37°C, 98.6°F).
Immune system	The immune system is the system of specialized cells and organs that protect an organism from outside biological influences. When the immune system is functioning properly, it protects the body against bacteria and viral infections, destroying cancer cells and foreign substances.
Inhibitor	An inhibitor is a type of effector (biology) that decreases or prevents the rate of a chemical reaction. They are often called negative catalysts.
Fetus	Fetus refers to a developing human from the ninth week of gestation until birth; has all the major structures of an adult.
Hibernation	Hibernation is a state of regulated hypothermia, lasting several days or weeks, that allows animals to conserve energy during the winter. During hibernation animals slow their metabolism to a very low level, with body temperature and breathing rates lowered, gradually using up the body fat reserves stored during the warmer months.

Blood	Blood is a circulating tissue composed of fluid plasma and cells. The main function of blood is to supply nutrients (oxygen, glucose) and constitutional elements to tissues and to remove waste products.
Secretion	Secretion is the process of segregating, elaborating, and releasing chemicals from a cell, or a secreted chemical substance or amount of substance.
Hormone	A hormone is a chemical messenger from one cell to another. All multicellular organisms produce hormones. The best known hormones are those produced by endocrine glands of vertebrate animals, but hormones are produced by nearly every organ system and tissue type in a human or animal body. Hormone molecules are secreted directly into the bloodstream, they move by circulation or diffusion to their target cells, which may be nearby cells in the same tissue or cells of a distant organ of the body.
Target cell	A cell that responds to a regulatory signal, such as a hormone is a target cell.
Cell	The cell is the structural and functional unit of all living organisms, and is sometimes called the "building block of life."
Liver	The liver is an organ in vertebrates, including humans. It plays a major role in metabolism and has a number of functions in the body including drug detoxification, glycogen storage, and plasma protein synthesis. It also produces bile, which is important for digestion.
Tissue	Group of similar cells which perform a common function is called tissue.
Skin	Skin is an organ of the integumentary system composed of a layer of tissues that protect underlying muscles and organs.
Nerve	A nerve is an enclosed, cable-like bundle of nerve fibers or axons, which includes the glia that ensheath the axons in myelin.
Growth factor	Growth factor is a protein that acts as a signaling molecule between cells (like cytokines and hormones) that attaches to specific receptors on the surface of a target cell and promotes differentiation and maturation of these cells.
Evolution	In biology, evolution is the process by which novel traits arise in populations and are passed on from generation to generation. Its action over large stretches of time explains the origin of new species and ultimately the vast diversity of the biological world.
Invertebrate	Invertebrate is a term coined by Jean-Baptiste Lamarck to describe any animal without a spinal column. It therefore includes all animals except vertebrates (fish, reptiles, amphibians, birds and mammals).
Thyroxine	The thyroid hormone thyroxine is a tyrosine-based hormone produced by the thyroid gland. An important component in the synthesis is iodine. It acts on the body to increase the basal metabolic rate, affect protein synthesis and increase the body's sensitivity to catecholamines.
Prolactin	Prolactin is a hormone synthesised and secreted by lactotrope cells in the anterior pituitary gland. It is made up of 199 amino acids, and has a molecular weight of about 23,000 daltons and has many effects, the most significant of which is to stimulate the mammary glands to produce milk (lactation).
Reproduction	Biological reproduction is the biological process by which new individual organisms are produced. Reproduction is a fundamental feature of all known life; each individual organism exists as the result of reproduction by an antecedent.
Endocrine gland	An endocrine gland is one of a set of internal organs involved in the secretion of hormones into the blood. These glands are known as ductless, which means they do not have tubes inside them.

Gland	A gland is an organ in an animal's body that synthesizes a substance for release such as hormones, often into the bloodstream or into cavities inside the body or its outer surface.
Salivary gland	The salivary gland produces saliva, which keeps the mouth and other parts of the digestive system moist. It also helps break down carbohydrates and lubricates the passage of food down from the oro-pharynx to the esophagus to the stomach.
Incomplete metamorphosis	Incomplete metamorphosis refers to a type of development in certain insects, such as grasshoppers, in which the larvae resemble adults but are smaller and have different body proportions.
Metamorphosis	Metamorphosis is a process in biology by which an individual physically develops after birth or hatching, and involves significant change in form as well as growth and differentiation.
Molting	In animals, molting is the routine shedding off old feathers in birds, or of old skin in reptiles, or of old hairs in mammals. In arthropods, such as insects, arachnids and crustaceans, molting describes the shedding of its exoskeleton (which is often called its shell), typically to let it grow.
Brain	The part of the central nervous system involved in regulating and controlling body activity and interpreting information from the senses transmitted through the nervous system is referred to as the brain.
Nervous system	The nervous system of an animal coordinates the activity of the muscles, monitors the organs, constructs and processes input from the senses, and initiates actions.
Egg	An egg is the zygote, resulting from fertilization of the ovum. It nourishes and protects the embryo.
Complete metamorphosis	Complete metamorphosis is a process of development in which the immature form looks and acts differently from the adult; the stages of development are egg, larva, pupa, and adult.
Insect	An arthropod that usually has three body segments , three pairs of legs, and one or two pairs of wings is called an insect. They are the largest and (on land) most widely-distributed taxon within the phylum Arthropoda. They comprise the most diverse group of animals on the earth, with around 925,000 species described
Vertebrate	Vertebrate is a subphylum of chordates, specifically, those with backbones or spinal columns. They started to evolve about 530 million years ago during the Cambrian explosion, which is part of the Cambrian period.
Anterior pituitary	The anterior pituitary comprises the anterior lobe of the pituitary gland and is part of the endocrine system. Under the influence of the hypothalamus, the anterior pituitary produces and secretes several peptide hormones that regulate many physiological processes including stress, growth, and reproduction.
Posterior pituitary	The posterior pituitary gland comprises the posterior lobe of the pituitary gland and is part of the endocrine system. Despite its name, the posterior pituitary gland is not a gland, rather, it is largely a collection of axonal projections from the hypothalamus that terminate behind the anterior pituitary gland.
Hypothalamus	Located below the thalamus, the hypothalamus links the nervous system to the endocrine system by synthesizing and secreting neurohormones often called releasing hormones because they function by stimulating the secretion of hormones from the anterior pituitary gland.
Vesicle	In cell biology, a vesicle is a relatively small and enclosed compartment, separated from the cytosol by at least one lipid bilayer.
Antidiuretic hormone	Antidiuretic hormone is a hormone that is mainly released when the body is low on water; it causes the kidneys to save water by concentrating the urine and is also involved in the

creation of thirst. It is a peptide hormone produced by the hypothalamus, and stored in the posterior part of the pituitary gland.

Growth hormone	Growth hormone is a polypeptide hormone synthesised and secreted by the anterior pituitary gland which stimulates growth and cell reproduction in humans and other vertebrate animals.
Genetic engineering	Genetic engineering, genetic modification (GM), and the now-deprecated gene splicing are terms for the process of manipulating genes,usually outside the organism's normal reproductive process.
Gene	Gene refers to a discrete unit of hereditary information consisting of a specific nucleotide sequence in DNA . Most of the genes of a eukaryote are located in its chromosomal DNA; a few are carried by the DNA of mitochondria and chloroplasts.
Melanocyte-stimulating hormone	Melanocyte-stimulating hormone is a peptide hormone produced by cells in the intermediate lobe of the pituitary gland. It stimulates the production and release of melanin (melanogenesis) by melanocytes in skin and hair.
Mammal	Homeothermic vertebrate characterized especially by the presence of hair and mammary glands is a mammal.
Negative feedback	Negative feedback refers to a control mechanism in which a chemical reaction, metabolic pathway, or hormonesecreting gland is inhibited by the products of the reaction, pathway, or gland.
Thyroid	The thyroid is one of the larger endocrine glands in the body. It is located in the neck and produces hormones, principally thyroxine and triiodothyronine, that regulate the rate of metabolism and affect the growth and rate of function of many other systems in the body.
Amino acid	An amino acid is any molecule that contains both amino and carboxylic acid functional groups. They are the basic structural building units of proteins. They form short polymer chains called peptides or polypeptides which in turn form structures called proteins.
Acid	An acid is a water-soluble, sour-tasting chemical compound that when dissolved in water, gives a solution with a pH of less than 7.
Goiter	Goiter refers to an enlargement of the thyroid gland resulting from a dietary iodine deficiency.
Autoimmune disease	Disease that results when the immune system mistakenly attacks the body's own tissues is referred to as autoimmune disease.
Metabolism	Metabolism is the biochemical modification of chemical compounds in living organisms and cells. This includes the biosynthesis of complex organic molecules (anabolism) and their breakdown (catabolism).
Ion	Ion refers to an atom or molecule that has gained or lost one or more electrons, thus acquiring an electrical charge.
Parathyroid gland	One of four endocrine glands embedded in the surface of the thyroid gland that secrete parathyroid hormone is called a parathyroid gland.
Calcitonin	Calcitonin is a a 32 amino acid polypeptide hormone that is produced in humans primarily by the C cells of the thyroid, and in many other animals in the ultimobranchial body.
Pancreas	The pancreas is a retroperitoneal organ that serves two functions: exocrine - it produces pancreatic juice containing digestive enzymes, and endocrine - it produces several important hormones, namely insulin.
Diabetes mellitus	Diabetes mellitus is a medical disorder characterized by varying or persistent hyperglycemia (elevated blood sugar levels), especially after eating. All types of diabetes mellitus share

Go to **Cram101.com** for the Practice Tests for this Chapter.

similar symptoms and complications at advanced stages.

Insulin	Insulin is a polypeptide hormone that regulates carbohydrate metabolism. Apart from being the primary effector in carbohydrate homeostasis, it also has a substantial effect on small vessel muscle tone, controls storage and release of fat (triglycerides) and cellular uptake of both amino acids and some electrolytes.
Receptor	A receptor is a protein on the cell membrane or within the cytoplasm or cell nucleus that binds to a specific molecule (a ligand), such as a neurotransmitter, hormone, or other substance, and initiates the cellular response to the ligand. Receptor, in immunology, the region of an antibody which shows recognition of an antigen.
Glucose	Glucose, a simple monosaccharide sugar, is one of the most important carbohydrates and is used as a source of energy in animals and plants. Glucose is one of the main products of photosynthesis and starts respiration.
Glycogen	Glycogen refers to a complex, extensively branched polysaccharide of many glucose monomers; serves as an energy-storage molecule in liver and muscle cells.
Adrenal	In mammals, the adrenal glands are the triangle-shaped endocrine glands that sit atop the kidneys. They are chiefly responsible for regulating the stress response through the synthesis of corticosteroids and catecholamines, including cortisol and adrenaline.
Adrenal medulla	Composed mainly of hormone-producing chromaffin cells, the adrenal medulla is the principal site of the conversion of the amino acid tyrosine into the catecholamines epinephrine and norepinephrine.
Medulla	Medulla in general means the inner part, and derives from the Latin word for 'marrow'. In medicine it is contrasted to the cortex.
Norepinephrine	Norepinephrine is a catecholamine and a phenethylamine with chemical formula $C_8H_{11}NO_3$. It is released from the adrenal glands as a hormone into the blood, but it is also a neurotransmitter in the nervous system where it is released from noradrenergic neurons during synaptic transmission.
Epinephrine	Epinephrine is a hormone and a neurotransmitter. Epinephrine plays a central role in the short-term stress reaction—the physiological response to threatening or exciting conditions (fight-or-flight response). It is secreted by the adrenal medulla.
Adrenal cortex	Situated along the perimeter of the adrenal gland, the adrenal cortex mediates the stress response through the production of mineralocorticoids and glucocorticoids, including aldosterone and cortisol respectively. It is also a secondary site of androgen synthesis.
Cortex	In anatomy and zoology the cortex is the outermost or superficial layer of an organ or the outer portion of the stem or root of a plant.
Mineralocort-coid	Mineralocorticoid is a class of steroids characterized by their similarity to aldosterone and their influence on salt and water metabolism.
Glucocorticoid	Glucocorticoid is a class of steroid hormones characterized by the ability to bind with the cortisol receptor and trigger similar effects. They are distinguished from mineralocorticoids and sex steroids by the specific receptors, target cells, and effects.
Stimulus	Stimulus in a nervous system, a factor that triggers sensory transduction.
Androgen	Androgen is the generic term for any natural or synthetic compound, usually a steroid hormone, that stimulates or controls the development and maintenance of masculine characteristics in vertebrates by binding to androgen receptors.
Estrogen	Estrogen is a steroid that functions as the primary female sex hormone. While present in both men and women, they are found in women in significantly higher quantities.

Go to **Cram101.com** for the Practice Tests for this Chapter.

273

Embryo	Embryo refers to a developing stage of a multicellular organism. In humans, the stage in the development of offspring from the first division of the zygote until body structures begin to appear, about the ninth week of gestation.
Y chromosome	Male sex chromosome that carries genes involved in sex determination is referred to as the Y chromosome. It contains the genes that cause testis development, thus determining maleness.
Chromosome	A chromosome is, minimally, a very long, continuous piece of DNA, which contains many genes, regulatory elements and other intervening nucleotide sequences.
Hermaphrodite	Hermaphrodite refers to an organism of a species whose members possess both male and female sexual organs during their lives. In many species, hermaphroditism is a normal part of the life-cycle. Generally, hermaphroditism occurs in the invertebrates, although it occurs in a fair number of fish, and to a lesser degree in other vertebrates.
Puberty	A time in the life of a developing individual characterized by the increasing production of sex hormones, which cause it to reach sexual maturity is called puberty.
Testes	The testes are the male generative glands in animals. Male mammals have two testes, which are often contained within an extension of the abdomen called the scrotum.
Muscle	Muscle is a contractile form of tissue. It is one of the four major tissue types, the other three being epithelium, connective tissue and nervous tissue. Muscle contraction is used to move parts of the body, as well as to move substances within the body.
Uterus	The uterus is the major female reproductive organ of most mammals. One end, the cervix, opens into the vagina; the other is connected on both sides to the fallopian tubes. The main function is to accept a fertilized ovum which becomes implanted into the endometrium, and derives nourishment from blood vessels which develop exclusively for this purpose.
Pineal gland	The pineal gland is a small endocrine gland in the brain. It is located near the center of the brain, between the two hemispheres, tucked in a groove where the two rounded thalamic bodies join.
Species	Group of similarly constructed organisms capable of interbreeding and producing fertile offspring is a species.
Melatonin	Melatonin, 5-methoxy-N-acetyltryptamine, is a hormone produced by pinealocytes in the pineal gland (located in the brain) and also by the retina and GI tract. Production of melatonin by the pineal gland is stimulated by darkness and inhibited by light.
Blood pressure	Blood pressure is the pressure exerted by the blood on the walls of the blood vessels.
Transduction	In physiology, transduction is transportation of a stimuli to the nervous system. In genetics, transduction is the transfer of viral, bacterial, or both bacterial and viral DNA from one cell to another via bacteriophage.
Molecule	A molecule is the smallest particle of a pure chemical substance that still retains its chemical composition and properties.
Lipid	Lipid is one class of aliphatic hydrocarbon-containing organic compounds essential for the structure and function of living cells. They are characterized by being water-insoluble but soluble in nonpolar organic solvents.
Domain	In biology, a domain is the top-level grouping of organisms in scientific classification.
Positive feedback	Mechanism of homeostatic response in which the output intensifies and increases the likelihood of response, instead of countering it and canceling it is called positive feedback.
Threshold	Electrical potential level at which an action potential or nerve impulse is produced is

Go to **Cram101.com** for the Practice Tests for this Chapter.

275

	called threshold.
Enzyme	An enzyme is a protein that catalyzes, or speeds up, a chemical reaction. They are essential to sustain life because most chemical reactions in biological cells would occur too slowly, or would lead to different products, without them.
Cortisol	Cortisol is a corticosteroid hormone that is involved in the response to stress; it increases blood pressure and blood sugar levels and suppresses the immune system.
Carrier protein	Protein molecule that combines with a substance and transports it through the plasma membrane is called a carrier protein.
Protein	A protein is a complex, high-molecular-weight organic compound that consists of amino acids joined by peptide bonds. They are essential to the structure and function of all living cells and viruses. Many are enzymes or subunits of enzymes.

Go to **Cram101.com** for the Practice Tests for this Chapter.

Species	Group of similarly constructed organisms capable of interbreeding and producing fertile offspring is a species.
Haploid	Haploid cells bear one copy of each chromosome.
Reproduction	Biological reproduction is the biological process by which new individual organisms are produced. Reproduction is a fundamental feature of all known life; each individual organism exists as the result of reproduction by an antecedent.
Population	Group of organisms of the same species occupying a certain area and sharing a common gene pool is referred to as population.
Multicellular	Multicellular organisms are those organisms consisting of more than one cell, and having differentiated cells that perform specialized functions. Most life that can be seen with the naked eye is multicellular, as are all animals (i.e. members of the kingdom Animalia) and plants (i.e. members of the kingdom Plantae).
Sponge	An invertebrates that consist of a complex aggregation of cells, including collar cells, and has a skeleton of fibers and/or spicules is a sponge. They are primitive, sessile, mostly marine, waterdwelling filter feeders that pump water through their matrix to filter out particulates of food matter.
Asexual reproduction	Asexual reproduction refers to reproduction that does not involve the fusion of haploid sex cells. The parent body may divide and new parts regenerate, or a new, smaller individual may form as an attachment to the parent, to drop off when complete.
Diploid	Diploid cells have two copies (homologs) of each chromosome (both sex- and non-sex determining chromosomes), usually one from the mother and one from the father. Most somatic cells (body cells) of complex organisms are diploid.
Meiosis	In biology, meiosis is the process that transforms one diploid cell into four haploid cells in eukaryotes in order to redistribute the diploid's cell's genome. Meiosis forms the basis of sexual reproduction and can only occur in eukaryotes.
Fertilization	Fertilization is fusion of gametes to form a new organism. In animals, the process involves a sperm fusing with an ovum, which eventually leads to the development of an embryo.
Testes	The testes are the male generative glands in animals. Male mammals have two testes, which are often contained within an extension of the abdomen called the scrotum.
Sperm	Sperm refers to the male sex cell with three distinct parts at maturity: head, middle piece, and tail.
Embryo	Embryo refers to a developing stage of a multicellular organism. In humans, the stage in the development of offspring from the first division of the zygote until body structures begin to appear, about the ninth week of gestation.
Gametogenesis	Gametogenesis is the creation of gametes by meiotic division of gametocytes into various gametes. Males and females of a species that reproduces sexually have different forms of gametogenesis.
X chromosome	The X chromosome is the female sex chromosome that carries genes involved in sex determination. Females have two X chromosomes, while males have one X and one Y chromosome.
Chromosome	A chromosome is, minimally, a very long, continuous piece of DNA, which contains many genes, regulatory elements and other intervening nucleotide sequences.
Y chromosome	Male sex chromosome that carries genes involved in sex determination is referred to as the Y chromosome. It contains the genes that cause testis development, thus determining maleness.
Gene	Gene refers to a discrete unit of hereditary information consisting of a specific nucleotide

Go to **Cram101.com** for the Practice Tests for this Chapter.

Go to **Cram101.com** for the Practice Tests for this Chapter.
And, **NEVER** highlight a book again!

sequence in DNA . Most of the genes of a eukaryote are located in its chromosomal DNA; a few are carried by the DNA of mitochondria and chloroplasts.

Egg
An egg is the zygote, resulting from fertilization of the ovum. It nourishes and protects the embryo.

Primary oocyte
The diploid cell of the ovary that begins to undergo the first meiotic division in the process of oogenesis is referred to as a primary oocyte.

Oocyte
An oocyte is a female gametocyte. Such that an oocyte is large and essentially stationary. The oocyte becomes functional when a lala (male gametocyte) attaches to it, thus allowing the meiosis of the secondary oocyte to occur.

Cell
The cell is the structural and functional unit of all living organisms, and is sometimes called the "building block of life."

Oogenesis
The formation of ova within the ovarian follicle of the ovary is referred to as oogenesis.

Invertebrate
Invertebrate is a term coined by Jean-Baptiste Lamarck to describe any animal without a spinal column. It therefore includes all animals except vertebrates (fish, reptiles, amphibians, birds and mammals).

Hermaphrodite
Hermaphrodite refers to an organism of a species whose members possess both male and female sexual organs during their lives. In many species, hermaphroditism is a normal part of the life-cycle. Generally, hermaphroditism occurs in the invertebrates, although it occurs in a fair number of fish, and to a lesser degree in other vertebrates.

Host
Host is an organism that harbors a parasite, mutual partner, or commensal partner; or a cell infected by a virus.

Reproductive system
A reproductive system is the ensembles and interactions of organs and or substances within an organism that stricly pertain to reproduction. As an example, this would include in the case of female mammals, the hormone estrogen, the womb and eggs but not the breast.

Organ
Organ refers to a structure consisting of several tissues adapted as a group to perform specific functions.

Spermatophore
In a variation on internal fertilization in some animals, the males package their sperm in a container that can be inserted into the female reproductive tract that is called a spermatophore.

Tentacle
An elongate, extensible projection of the body of cnidarians and cephalopod mollusks that may be used for grasping, stinging, and immobilizing prey, and for locomotion is called tentacle.

Insect
An arthropod that usually has three body segments , three pairs of legs, and one or two pairs of wings is called an insect. They are the largest and (on land) most widely-distributed taxon within the phylum Arthropoda. They comprise the most diverse group of animals on the earth, with around 925,000 species described

Vertebrate
Vertebrate is a subphylum of chordates, specifically, those with backbones or spinal columns. They started to evolve about 530 million years ago during the Cambrian explosion, which is part of the Cambrian period.

Internal fertilization
Internal fertilization is a form of animal fertilization of an ovum by spermatozoon within the body of an inseminated animal, whether female or hermaphroditic.

Blood
Blood is a circulating tissue composed of fluid plasma and cells. The main function of blood is to supply nutrients (oxygen, glucose) and constitutional elements to tissues and to remove waste products.

Oviparous
In some animals, an egg is the zygote, resulting from fertilization of the ovum. It nourishes

Go to **Cram101.com** for the Practice Tests for this Chapter.

and protects the embryo. Oviparous animals are animals that lay eggs, with little or no other development within the mother.

Viviparous	Viviparous refers to reproduction in which eggs develop within the mother's body and young are born free-living.
Nutrition	Nutrition refers to collectively, the processes involved in taking in, assimilating, and utilizing nutrients.
Adaptation	A biological adaptation is an anatomical structure, physiological process or behavioral trait of an organism that has evolved over a period of time by the process of natural selection such that it increases the expected long-term reproductive success of the organism.
Uterus	The uterus is the major female reproductive organ of most mammals. One end, the cervix, opens into the vagina; the other is connected on both sides to the fallopian tubes. The main function is to accept a fertilized ovum which becomes implanted into the endometrium, and derives nourishment from blood vessels which develop exclusively for this purpose.
Marsupium	The external pouch on the abdomen of a female marsupial is called marsupium.
Semen	Semen is a fluid that contains spermatozoa. It is secreted by the gonads (sexual glands) of male or hermaphroditic animals including humans for fertilization of female ova. Semen discharged by an animal or human is known as ejaculate, and the process of discharge is called ejaculation.
Scrotum	In some male mammals the scrotum is an external bag of skin and muscle containing the testicles. It is an extension of the abdomen, and is located between the penis and anus.
Seminiferous tubule	Highly coiled duct within the male testes that produces and transports sperm is the seminiferous tubule.
Nucleus	In cell biology, the nucleus is found in all eukaryotic cells that contains most of the cell's genetic material. The nucleus has two primary functions: to control chemical reactions within the cytoplasm and to store information needed for cellular division.
Acrosome	In the spermatozoa of many animals, the acrosome is an organelle that develops over the anterior half of the spermatozoon's head. It alters a patch of pre-existing sperm plasma membrane so that it can fuse with the egg plasma membrane.
Epididymis	The epididymis is part of the human male reproductive system and is present in all male mammals. It is a narrow, tightly-coiled tube connecting the efferent ducts from the rear of each testicle to its vas deferens.
Urethra	In anatomy, the urethra is a tube which connects the urinary bladder to the outside of the body. The urethra has an excretory function in both sexes, to pass urine to the outside, and also a reproductive function in the male, as a passage for sperm.
Penis	The penis is the male reproductive organ and for mammals additionally serves as the external male organ of urination.
Skin	Skin is an organ of the integumentary system composed of a layer of tissues that protect underlying muscles and organs.
Autonomic nervous system	The autonomic nervous system is the part of the nervous system that is not consciously controlled. It is commonly divided into two usually antagonistic subsystems: the sympathetic and parasympathetic nervous system.
Nervous system	The nervous system of an animal coordinates the activity of the muscles, monitors the organs, constructs and processes input from the senses, and initiates actions.
Tissue	Group of similar cells which perform a common function is called tissue.

Go to **Cram101.com** for the Practice Tests for this Chapter.

Blood pressure	Blood pressure is the pressure exerted by the blood on the walls of the blood vessels.
Prostate	The prostate is a gland that is part of male mammalian sex organs. Its main function is to secrete and store a clear, slightly basic fluid that is part of semen. The prostate differs considerably between species anatomically, chemically and physiologically.
Gland	A gland is an organ in an animal's body that synthesizes a substance for release such as hormones, often into the bloodstream or into cavities inside the body or its outer surface.
Cancer	Cancer is a class of diseases or disorders characterized by uncontrolled division of cells and the ability of these cells to invade other tissues, either by direct growth into adjacent tissue through invasion or by implantation into distant sites by metastasis.
Testosterone	Testosterone is a steroid hormone from the androgen group. Testosterone is secreted in the testes of men and the ovaries of women. It is the principal male sex hormone and the "original" anabolic steroid. In both males and females, it plays key roles in health and well-being.
Ovary	In the flowering plants, an ovary is a part of the female reproductive organ of the flower or gynoecium.
Oviduct	In oviparous animals (those that lay eggs), the passage from the ovaries to the outside of the body is known as the oviduct. The eggs travel along the oviduct.
Vagina	The vagina is the tubular tract leading from the uterus to the exterior of the body in female placental mammals and marsupials, or to the cloaca in female birds, monotremes, and some reptiles. Female insects and other invertebrates also have a vagina, which is the terminal part of the oviduct.
Labia minora	The labia minora are two soft folds of skin within the labia majora and to either side of the opening of the vagina.
Clitoris	The clitoris is a sexual organ in the body of female mammals. The visible knob-like portion is located near the anterior junction of the labia minora, above the opening of the vagina. Unlike its male counterpart, the penis,the clitoris has no urethra, is not involved in urination, and its sole function is to induce sexual pleasure.
Hymen	The hymen is a ring of tissue around the vaginal opening. Although many people believe that the hymen completely occludes the vaginal opening in human females, this is quite rare. The hymen has great symbolic significance as an indicator of a woman's virginity.
Blastocyst	A mammalian embryo made up of a hollow ball of cells that results from cleavage and that implants in the mother's endometrium is called blastocyst.
Endometrium	The endometrium is the inner uterine membrane in mammals which is developed in preparation for the implantation of a fertilized egg upon its arrival into the uterus.
Placenta	The placenta is an organ present only in female placental mammals during gestation. It is composed of two parts, one genetically and biologically part of the fetus, the other part of the mother. It is implanted in the wall of the uterus, where it receives nutrients and oxygen from the mother's blood and passes out waste.
Ovarian cycle	Hormonally synchronized cyclical events in the mammalian ovary, culminating in ovulation is called the ovarian cycle.
Puberty	A time in the life of a developing individual characterized by the increasing production of sex hormones, which cause it to reach sexual maturity is called puberty.
Follicle	Follicle refers to a cluster of cells surrounding, protecting, and nourishing a developing egg cell in the ovary; also secretes estrogen. In botany, a follicle is a type of simple dry fruit produced by certain flowering plants. It is regarded as one the most primitive types of

Go to **Cram101.com** for the Practice Tests for this Chapter.

285

fruits, and derives from a simple pistil or carpel.

Ovulation	Ovulation is the process in the menstrual cycle by which a mature ovarian follicle ruptures and discharges an ovum (also known as an oocyte, female gamete, or casually, an egg) that participates in reproduction.
Uterine cycle	Monthly occurring changes in the characteristics of the uterine lining preparing it for implantation of the blastocyst is the uterine cycle. Another term for menstrual cycle.
Anterior pituitary	The anterior pituitary comprises the anterior lobe of the pituitary gland and is part of the endocrine system. Under the influence of the hypothalamus, the anterior pituitary produces and secretes several peptide hormones that regulate many physiological processes including stress, growth, and reproduction.
Hypothalamus	Located below the thalamus, the hypothalamus links the nervous system to the endocrine system by synthesizing and secreting neurohormones often called releasing hormones because they function by stimulating the secretion of hormones from the anterior pituitary gland.
Estrogen	Estrogen is a steroid that functions as the primary female sex hormone. While present in both men and women, they are found in women in significantly higher quantities.
Negative feedback	Negative feedback refers to a control mechanism in which a chemical reaction, metabolic pathway, or hormonesecreting gland is inhibited by the products of the reaction, pathway, or gland.
Progesterone	Progesterone is a C-21 steroid hormone involved in the female menstrual cycle, pregnancy (supports gestation) and embryogenesis of humans and other species.
Biology	Biology is the branch of science dealing with the study of life. It is concerned with the characteristics, classification, and behaviors of organisms, how species come into existence, and the interactions they have with each other and with the environment.
Orgasm	Orgasm refers to rhythmic contractions of the reproductive structures, accompanied by extreme pleasure, at the peak of sexual excitement in both sexes; includes ejaculation by the male.
Resolution phase	The resolution phase occurs after orgasm and allows the muscles to relax, blood pressure to drop and the body to slow down from its excited state.
Nerve	A nerve is an enclosed, cable-like bundle of nerve fibers or axons, which includes the glia that ensheath the axons in myelin.
Basal body	A basal body is an organelle formed from a centriole, a short cylindrical array of microtubules. It is found at the base of a eukaryotic cell cilium or flagellum and serves as a nucleation site for the growth of the axoneme microtubules.
Ovum	An ovum is a female sex cell or gamete. It is a mature egg cell released during ovulation from an ovary.
Barrier method	A barrier method refers to methods of preventing pregnancy by forming an impenetrable barrier between sexual partners.
Condom	Sheath used to cover the penis during sexual intercourse is referred to as condom.
Syphilis	Syphilis refers to a sexually transmitted bacterial infection of the reproductive organs; if untreated, can damage the nervous and circulatory systems.
Diaphragm	The diaphragm is a shelf of muscle extending across the bottom of the ribcage. It is critically important in respiration: in order to draw air into the lungs, the diaphragm contracts, thus enlarging the thoracic cavity and reducing intra-thoracic pressure.
Implantation	Implantation refers to attachment and penetration of the embryo into the lining of the uterus.

Sterilization	Sterilization is the elimination of all transmissible agents (such as bacteria, prions and viruses) from a surface or piece of equipment. This is different from disinfection, where only organisms that can cause disease are removed by a disinfectant.
Tubal ligation	Tubal ligation refers to a means of sterilization in which a woman's two oviducts are tied closed to prevent eggs from reaching the uterus. A segment of each oviduct is removed.
Ligation	Ligation refers to enzymatically catalyzed formation of a phosphodiester bond that links two DNA molecules.
Abortion	An abortion is the termination of a pregnancy associated with the death of an embryo or a fetus.
Fetus	Fetus refers to a developing human from the ninth week of gestation until birth; has all the major structures of an adult.
Cervix	The cervix is actually the lower, narrow portion of the uterus where it joins with the top end of the vagina. It is cylindrical or conical in shape and protrudes through the upper anterior vaginal wall.
Spermatogenesis	Spermatogenesis refers to the creation, or genesis, of spermatozoa, which occurs in the male gonads.
In vitro fertilization	Uniting sperm and egg in a laboratory container, followed by the placement of a resulting early embryo in the mother's uterus is referred to as in vitro fertilization.
In vitro	In vitro is an experimental technique where the experiment is performed in a test tube, or generally outside a living organism or cell.
Gamete	A gamete is a specialized germ cell that unites with another gamete during fertilization in organisms that reproduce sexually. They are haploid cells; that is, they contain one complete set of chromosomes. When they unite they form a zygote—a cell having two complete sets of chromosomes and therefore diploid.
Copulation	Copulation is the union of the external sexual organs of two sexually reproducing animal organisms for insemination and for subsequent internal fertilization, which is fertilization of ova inside organisms.
Chlamydia	A sexually transmitted disease, caused by a bacterium, that causes inflammation of the urethra in males and of the urethra and cervix in females is referred to as chlamydia.
Virus	Obligate intracellular parasite of living cells consisting of an outer capsid and an inner core of nucleic acid is referred to as virus. The term virus usually refers to those particles that infect eukaryotes whilst the term bacteriophage or phage is used to describe those infecting prokaryotes.

Go to **Cram101.com** for the Practice Tests for this Chapter.

Vertebrate	Vertebrate is a subphylum of chordates, specifically, those with backbones or spinal columns. They started to evolve about 530 million years ago during the Cambrian explosion, which is part of the Cambrian period.
Egg	An egg is the zygote, resulting from fertilization of the ovum. It nourishes and protects the embryo.
Transcription	Transcription is the process through which a DNA sequence is enzymatically copied by an RNA polymerase to produce a complementary RNA. Or, in other words, the transfer of genetic information from DNA into RNA.
Species	Group of similarly constructed organisms capable of interbreeding and producing fertile offspring is a species.
Sperm	Sperm refers to the male sex cell with three distinct parts at maturity: head, middle piece, and tail.
Plasma	In physics and chemistry, a plasma is an ionized gas, and is usually considered to be a distinct phase of matter. "Ionized" in this case means that at least one electron has been dissociated from a significant fraction of the molecules.
Plasma membrane	Membrane surrounding the cytoplasm that consists of a phospholipid bilayer with embedded proteins is referred to as plasma membrane.
Acrosome	In the spermatozoa of many animals, the acrosome is an organelle that develops over the anterior half of the spermatozoon's head. It alters a patch of pre-existing sperm plasma membrane so that it can fuse with the egg plasma membrane.
Actin	Actin is a globular protein that polymerizes helically forming filaments, which like the other two components of the cellular cytoskeleton form a three-dimensional network inside an eukaryotic cell. They provide mechanical support for the cell, determine the cell shape, enable cell movements .
Glycoprotein	A macromolecule consisting of one or more polypeptides linked to short chains of sugars is called glycoprotein.
Zona pellucida	The zona pellucida is a glycoprotein matrix surrounding the plasma membrane of an oocyte. This structure binds spermatozoa, and is required to initiate the acrosome reaction.
Hybrid	Hybrid refers to the offspring of parents of two different species or of two different varieties of one species; the offspring of two parents that differ in one or more inherited traits; an individual that is heterozygous for one or more pair of genes.
Fertilization	Fertilization is fusion of gametes to form a new organism. In animals, the process involves a sperm fusing with an ovum, which eventually leads to the development of an embryo.
Enzyme	An enzyme is a protein that catalyzes, or speeds up, a chemical reaction. They are essential to sustain life because most chemical reactions in biological cells would occur too slowly, or would lead to different products, without them.
Endoplasmic reticulum	The endoplasmic reticulum is an organelle found in all eukaryotic cells. It modifies proteins, makes macromolecules, and transfers substances throughout the cell.
Cytoplasm	Cytoplasm refers to everything inside a cell between the plasma membrane and the nucleus; consists of a semifluid medium and organelles.
Haploid	Haploid cells bear one copy of each chromosome.
Gray crescent	Gray area that appears in an amphibian egg after being fertilized by the sperm, thought to contain chemical signals that turn on the genes that control development is referred to as gray crescent.

Go to **Cram101.com** for the Practice Tests for this Chapter.

Zygote	Diploid cell formed by the union of sperm and egg is referred to as zygote.
Microtubule	Microtubule is a protein structure found within cells, one of the components of the cytoskeleton. They have diameter of ~ 24 nm and varying length from several micrometers to possible millimeters in axons of nerve cells. They serve as structural components within cells and are involved in many cellular processes including mitosis, cytokinesis, and vesicular transport.
Cell	The cell is the structural and functional unit of all living organisms, and is sometimes called the "building block of life."
Cleavage	Cleavage refers to cytokinesis in animal cells and in some protists, characterized by pinching in of the plasma membrane.
Embryo	Embryo refers to a developing stage of a multicellular organism. In humans, the stage in the development of offspring from the first division of the zygote until body structures begin to appear, about the ninth week of gestation.
Blastula	The blastula is an early stage of embryonic development in animals. It is produced by cleavage of a fertilized ovum and consisting of a spherical layer of cells surrounding a fluid-filled cavity.
Yolk	Dense nutrient material that is present in the egg of a bird or reptile is referred to as yolk.
Insect	An arthropod that usually has three body segments , three pairs of legs, and one or two pairs of wings is called an insect. They are the largest and (on land) most widely-distributed taxon within the phylum Arthropoda. They comprise the most diverse group of animals on the earth, with around 925,000 species described
Trophoblast	The outer membrane surrounding the embryo in mammals is called the trophoblast.
Blastocyst	A mammalian embryo made up of a hollow ball of cells that results from cleavage and that implants in the mother's endometrium is called blastocyst.
Implantation	Implantation refers to attachment and penetration of the embryo into the lining of the uterus.
Blastocoel	A blastocoel is the central region of a blastula. It is filled with fluid. A blastocoele forms during embryogenesis when a zygote divides into many cells through mitosis.
Tissue	Group of similar cells which perform a common function is called tissue.
Gastrulation	The phase of embryonic development that transforms the blastula into a gastrula. Gastrulation adds more cells to the embryo and sorts the cells into distinct cell layers.
Germ layer	A germ layer is a collection of cells, formed during early embryonic development. It will eventually give rise to all of an animal's tissues and organs through a process called organogenesis.
Archenteron	The archenteron is an indentation that forms early on in a developing blastula. The archenteron will go on to develop into either the animal's anus or mouth.
Endoderm	Cells migrating inward along the archenteron form the inner layer of the gastrula, which develops into the endoderm.
Dorsal	In anatomy, the dorsal is the side in which the backbone is located. This is usually the top of an animal, although in humans it refers to the back.
Blastopore	A blastopore is an opening into the archenteron during the embryonic stages of an organism. The distinction between protostomes and deuterostomes is based on the direction in which the mouth (stoma) develops in relation to the blastopore.

Go to **Cram101.com** for the Practice Tests for this Chapter.

Epidermis	Epidermis is the outermost layer of the skin. It forms the waterproof, protective wrap over the body's surface and is made up of stratified squamous epithelium with an underlying basement membrane. It contains no blood vessels, and is nourished by diffusion from the dermis. In plants, the outermost layer of cells covering the leaves and young parts of a plant is the epidermis.
Molecule	A molecule is the smallest particle of a pure chemical substance that still retains its chemical composition and properties.
Primitive streak	Primitive streak is a faint white trace at the caudal end of the embryonic disc, formed by the movement of cells at the beginning of the mesoderm formation. It provides the earliest indication of the embryonic axis.
Node	In botany, a node is the place on a stem where a leaf is attached.
Amnion	The amnion is a membranous sac which surrounds and protects the embryo. It is developed in reptiles, birds, and mammals, which are hence called "Amniota;" but not in amphibia and fishes, which are consequently termed "Anamnia."
Neurulation	The elaboration of a notochord and a dorsal nerve cord that marks the evolution of the chordates is referred to as neurulation.
Neural tube	The neural tube is the embryonal structure that gives rise to the brain and spinal cord. The neural tube is derived from a thickened area of ectoderm, the neural plate. The process of formation of the neural tube is called neurulation.
Nerve	A nerve is an enclosed, cable-like bundle of nerve fibers or axons, which includes the glia that ensheath the axons in myelin.
Homeobox	A homeobox is a DNA sequence found within genes that are involved in the regulation of development of animals, fungi and plants.
Spinal cord	The spinal cord is a part of the vertebrate nervous system that is enclosed in and protected by the vertebral column (it passes through the spinal canal). It consists of nerve cells. The spinal cord carries sensory signals and motor innervation to most of the skeletal muscles in the body.
Yolk sac	The yolk sac is the first element seen in the gestational sac during pregnancy, usually at 5 weeks gestation. It is filled with fluid, the vitelline fluid, which possibly may be utilized for the nourishment of the embryo during the earlier stages of its existence.
Allantois	Allantois is a part of a developing animal embryo. This sac-like structure is primarily involved in respiration and excretion, and is webbed with blood vessels. It is primarily found in the blastocyst stage of early embryological development, and its purpose is to collect liquid waste from the embryo.
Chorion	In animals, the outermost extraembryonic membrane, which becomes the mammalian embryo's part of the placenta is referred to as the chorion.
Umbilical cord	A structure containing arteries and veins that connects a developing embryo to the placenta of the mother is an umbilical cord.
Amniocentesis	Amniocentesis is a procedure used for prenatal diagnosis, in which a small amount of amniotic fluid is extracted from the amnion around a developing fetus. It is usually offered when there may be an increased risk for genetic conditions (i.e. Down syndrome, sickle-cell disease, cystic fibrosis, etc) in the pregnancy.
Fetus	Fetus refers to a developing human from the ninth week of gestation until birth; has all the major structures of an adult.
Chorionic	Chorionic villus sampling is a form of prenatal diagnosis to determine genetic abnormalities

villus sampling	in the fetus. It entails getting a sample of the chorionic villus (placental tissue) and testing it.
Villus	Villus refers to a fingerlike projection of the inner surface of the small intestine. A fingerlike projection of the chorion of the mammalian placenta. Large numbers of villus increase the surface areas of these organs.
Trimester	In human development, one of three 3-mnonth-long periods of pregnancy is called trimester.
Gestation	Gestation refers to pregnancy; the state of carrying developing young within the female reproductive tract.
Estrogen	Estrogen is a steroid that functions as the primary female sex hormone. While present in both men and women, they are found in women in significantly higher quantities.
Digestive system	The organ system that ingests food, breaks it down into smaller chemical units, and absorbs the nutrient molecules is referred to as the digestive system.
Uterus	The uterus is the major female reproductive organ of most mammals. One end, the cervix, opens into the vagina; the other is connected on both sides to the fallopian tubes. The main function is to accept a fertilized ovum which becomes implanted into the endometrium, and derives nourishment from blood vessels which develop exclusively for this purpose.
Secretion	Secretion is the process of segregating, elaborating, and releasing chemicals from a cell, or a secreted chemical substance or amount of substance.

Nervous system	The nervous system of an animal coordinates the activity of the muscles, monitors the organs, constructs and processes input from the senses, and initiates actions.
Organ	Organ refers to a structure consisting of several tissues adapted as a group to perform specific functions.
Multicellular	Multicellular organisms are those organisms consisting of more than one cell, and having differentiated cells that perform specialized functions. Most life that can be seen with the naked eye is multicellular, as are all animals (i.e. members of the kingdom Animalia) and plants (i.e. members of the kingdom Plantae).
Brain	The part of the central nervous system involved in regulating and controlling body activity and interpreting information from the senses transmitted through the nervous system is referred to as the brain.
Vertebrate	Vertebrate is a subphylum of chordates, specifically, those with backbones or spinal columns. They started to evolve about 530 million years ago during the Cambrian explosion, which is part of the Cambrian period.
Cell body	The part of a cell, such as a neuron, that houses the nucleus is the cell body.
Cell	The cell is the structural and functional unit of all living organisms, and is sometimes called the "building block of life."
Axon	An axon is a long slender projection of a nerve cell, or neuron, which conducts electrical impulses away from the neuron's cell body or soma. They are in effect the primary transmission lines of the nervous system, and as bundles they help make up nerves.
Dendrite	A dendrite is a slender, typically branched projection of a neuron, which conducts the electrical stimulation received from other cells to and from the cell body or soma of the neuron from which it projects. This stimulation arrives through synapses, which are located at various points throughout the dendritic arbor.
Target cell	A cell that responds to a regulatory signal, such as a hormone is a target cell.
Synapse	A junction, or relay point, between two neurons, or between a neuron and an effector cell. Electrical and chemical signals are relayed from one cell to another at a synapse.
Nerve impulse	Action potential traveling along a neuron is a nerve impulse.
Nerve	A nerve is an enclosed, cable-like bundle of nerve fibers or axons, which includes the glia that ensheath the axons in myelin.
Synaptic cleft	A narrow gap separating the synaptic knob of a transmitting neuron from a receiving neuron or an effector cell is a synaptic cleft.
Neuron	The neuron is a major class of cells in the nervous system. In vertebrates, they are found in the brain, the spinal cord and in the nerves and ganglia of the peripheral nervous system, and their primary role is to process and transmit neural information.
Peripheral nervous system	The peripheral nervous system consists of the nerves and neurons that reside or extend outside the central nervous system--to serve the limbs and organs. The peripheral nervous system is divided into the somatic nervous system and the autonomic nervous system.
Myelin	Myelin is an electrically insulating fatty layer that surrounds the axons of many neurons, especially those in the peripheral nervous system. It is an outgrowth of glial cells: Schwann cells supply the myelin for peripheral neurons while oligodendrocytes supply it to those of the central nervous system.
Blood-brain barrier	The blood-brain barrier is a membrane that controls the passage of substances from the blood into the central nervous system. It is a physical barrier between the blood vessels in the

Go to **Cram101.com** for the Practice Tests for this Chapter.

	central nervous system, and most parts of the central nervous system itself, that stops many substances from traveling across it.
Plasma	In physics and chemistry, a plasma is an ionized gas, and is usually considered to be a distinct phase of matter. "Ionized" in this case means that at least one electron has been dissociated from a significant fraction of the molecules.
Sensory neuron	Sensory neuron refers to nerve cell that transmits nerve impulses to the central nervous system after a sensory receptor has been stimulated.
Voltage	A measure of the electrical difference that exists between two different points or objects is called voltage.
Plasma membrane	Membrane surrounding the cytoplasm that consists of a phospholipid bilayer with embedded proteins is referred to as plasma membrane.
Resting potential	The resting potential of a cell is the membrane potential that would be maintained if there were no action potentials, synaptic potentials, or other active changes in the membrane potential.
Action potential	An action potential is a wave of electrical discharge that travels along the membrane of a cell. They communicate fast internal messages between tissues making them an essential feature of animal life at the microscopic level.
Electron	The electron is a light fundamental subatomic particle that carries a negative electric charge. The electron is a spin-1/2 lepton, does not participate in strong interactions and has no substructure.
Ion	Ion refers to an atom or molecule that has gained or lost one or more electrons, thus acquiring an electrical charge.
Lipid bilayer	A lipid bilayer is a membrane or zone of a membrane composed of lipid molecules (usually phospholipids). The lipid bilayer is a critical component of all biological membranes, including cell membranes, and is a prerequisite for cell-based organisms.
Lipid	Lipid is one class of aliphatic hydrocarbon-containing organic compounds essential for the structure and function of living cells. They are characterized by being water-insoluble but soluble in nonpolar organic solvents.
Diffusion	Diffusion refers to the spontaneous movement of particles of any kind from where they are more concentrated to where they are less concentrated.
Net movement	Net movement refers to movement in one direction minus the movement in the other.
Channel protein	Membrane transport protein that forms an aqueous pore in the membrane through which a specific solute, usually an ion, can pass is the channel protein.
Protein	A protein is a complex, high-molecular-weight organic compound that consists of amino acids joined by peptide bonds. They are essential to the structure and function of all living cells and viruses. Many are enzymes or subunits of enzymes.
Stimulus	Stimulus in a nervous system, a factor that triggers sensory transduction.
Invertebrate	Invertebrate is a term coined by Jean-Baptiste Lamarck to describe any animal without a spinal column. It therefore includes all animals except vertebrates (fish, reptiles, amphibians, birds and mammals).
Node	In botany, a node is the place on a stem where a leaf is attached.
Stem	Stem refers to that part of a plant's shoot system that supports the leaves and reproductive structures.

Go to **Cram101.com** for the Practice Tests for this Chapter.

301

Motor neuron	A nerve cell that conveys command signals from the central nervous system to effector cells, such as muscle cells or gland cells is a motor neuron.
Neurotransmitter	A neurotransmitter is a chemical that is used to relay, amplify and modulate electrical signals between a neuron and another cell.
Exocytosis	Exocytosis is the process by which a cell is able to release large biomolecules through its membrane. While in protozoa the exocytosis may serve the function of wasting unnecessary products, in multicellular organisms exocytosis serves signalling or regulatory function.
Acetylcholine	The chemical compound acetylcholine was the first neurotransmitter to be identified. It is a chemical transmitter in both the peripheral nervous system (PNS) and central nervous system (CNS) in many organisms including humans.
Enzyme	An enzyme is a protein that catalyzes, or speeds up, a chemical reaction. They are essential to sustain life because most chemical reactions in biological cells would occur too slowly, or would lead to different products, without them.
Peptide	Peptide is the family of molecules formed from the linking, in a defined order, of various amino acids. The link between one amino acid residue and the next is an amide bond, and is sometimes referred to as a peptide bond.
Muscle	Muscle is a contractile form of tissue. It is one of the four major tissue types, the other three being epithelium, connective tissue and nervous tissue. Muscle contraction is used to move parts of the body, as well as to move substances within the body.
Receptor	A receptor is a protein on the cell membrane or within the cytoplasm or cell nucleus that binds to a specific molecule (a ligand), such as a neurotransmitter, hormone, or other substance, and initiates the cellular response to the ligand. Receptor, in immunology, the region of an antibody which shows recognition of an antigen.
Neuromuscular junction	A neuromuscular junction is the junction of the axon terminal of a motoneuron with the motor end plate, the highly-excitable region of muscle fiber plasma membrane responsible for initiation of action potentials across the muscle's surface.
Postsynaptic neuron	At a synapse, the nerve cell that changes its electrical potential in response to a chemical released by another cell is a postsynaptic neuron.
Central nervous system	The central nervous system comprized of the brain and spinal cord, represents the largest part of the nervous system. Together with the peripheral nervous system, it has a fundamental role in the control of behavior.
Smooth muscle	Smooth muscle is a type of non-striated muscle, found within the "walls" of hollow organs; such as blood vessels, the bladder, the uterus, and the gastrointestinal tract. Smooth muscle is used to move matter within the body, via contraction; it generally operates "involuntarily", without nerve stimulation.
Depolarization	Depolarization is a decrease in the absolute value of a cell's membrane potential.
Acetylcholin-sterase	Acetylcholinesterase is found primarily in the blood and neural synapses. It catalyzes the hydrolysis of the neurotransmitter acetylcholine into choline and acetic acid, a reaction necessary to allow a cholinergic neuron to return to its resting state after activation.
Active transport	Active transport is the mediated transport of biochemicals, and other atomic/molecular substances, across membranes. In this form of transport, molecules move against either an electrical or concentration gradient.

Species	Group of similarly constructed organisms capable of interbreeding and producing fertile offspring is a species.
Receptor	A receptor is a protein on the cell membrane or within the cytoplasm or cell nucleus that binds to a specific molecule (a ligand), such as a neurotransmitter, hormone, or other substance, and initiates the cellular response to the ligand. Receptor, in immunology, the region of an antibody which shows recognition of an antigen.
Nervous system	The nervous system of an animal coordinates the activity of the muscles, monitors the organs, constructs and processes input from the senses, and initiates actions.
Cortex	In anatomy and zoology the cortex is the outermost or superficial layer of an organ or the outer portion of the stem or root of a plant.
Skin	Skin is an organ of the integumentary system composed of a layer of tissues that protect underlying muscles and organs.
Brain	The part of the central nervous system involved in regulating and controlling body activity and interpreting information from the senses transmitted through the nervous system is referred to as the brain.
Photoreceptor	A photoreceptor is a specialized type of neuron that is capable of phototransduction. More specifically, the photoreceptor sends signals to other neurons by a change in its membrane potential when it absorbs photons.
Cell	The cell is the structural and functional unit of all living organisms, and is sometimes called the "building block of life."
Receptor protein	Protein located in the plasma membrane or within the cell that binds to a substance that alters some metabolic aspect of the cell is referred to as receptor protein. It will only link up with a substance that has a certain shape that allows it to bind to the receptor.
Protein	A protein is a complex, high-molecular-weight organic compound that consists of amino acids joined by peptide bonds. They are essential to the structure and function of all living cells and viruses. Many are enzymes or subunits of enzymes.
Axon	An axon is a long slender projection of a nerve cell, or neuron, which conducts electrical impulses away from the neuron's cell body or soma. They are in effect the primary transmission lines of the nervous system, and as bundles they help make up nerves.
Ion	Ion refers to an atom or molecule that has gained or lost one or more electrons, thus acquiring an electrical charge.
Stimulus	Stimulus in a nervous system, a factor that triggers sensory transduction.
Molecule	A molecule is the smallest particle of a pure chemical substance that still retains its chemical composition and properties.
Chemoreceptor	Chemoreceptor is a cell or group of cells that transduce a chemical signal into an action potential.
Olfaction	Olfaction, the sense of odor, is the detection of chemicals dissolved in air. In vertebrates smells are sensed by the olfactory epithelium located in the nasal cavity and first processed by the olfactory bulb in the olfactory system. In insects smells are sensed by sensilia located on the antenna and first processed by the antennal lobe.
Vertebrate	Vertebrate is a subphylum of chordates, specifically, those with backbones or spinal columns. They started to evolve about 530 million years ago during the Cambrian explosion, which is part of the Cambrian period.
Epithelium	Epithelium is a tissue composed of a layer of cells. Epithelium can be found lining internal

Go to **Cram101.com** for the Practice Tests for this Chapter.

(e.g. endothelium, which lines the inside of blood vessels) or external (e.g. skin) free surfaces of the body. Functions include secretion, absorption and protection.

Second messenger	An intermediary compound that couples extracellular signals to intracellular processes and also amplifies a hormonal signal is referred to as second messenger.
Monotreme	Monotreme is a mammal that lays eggs, instead of giving birth to live young like marsupials (Metatheria) and placental mammals (Eutheria).
Taste bud	A taste bud is a small structure on the upper surface of the tongue, soft palate, and epiglottis that provides information about the taste of food being eaten. The majority on the tongue sit on raized protrusions of the tongue surface called papillae.
Transduction	In physiology, transduction is transportation of a stimuli to the nervous system. In genetics, transduction is the transfer of viral, bacterial, or both bacterial and viral DNA from one cell to another via bacteriophage.
Sensory neuron	Sensory neuron refers to nerve cell that transmits nerve impulses to the central nervous system after a sensory receptor has been stimulated.
Neuron	The neuron is a major class of cells in the nervous system. In vertebrates, they are found in the brain, the spinal cord and in the nerves and ganglia of the peripheral nervous system, and their primary role is to process and transmit neural information.
Blood	Blood is a circulating tissue composed of fluid plasma and cells. The main function of blood is to supply nutrients (oxygen, glucose) and constitutional elements to tissues and to remove waste products.
Mechanoreceptor	A mechanoreceptor is a sensory receptor that responds to mechanical pressure or distortion. There are four main types: Pacinian corpuscles, Meissner's corpuscles, Merkel's discs, and the tympanic membrane.
Connective tissue	Connective tissue is any type of biological tissue with an extensive extracellular matrix and often serves to support, bind together, and protect organs.
Tissue	Group of similar cells which perform a common function is called tissue.
Plasma membrane	Membrane surrounding the cytoplasm that consists of a phospholipid bilayer with embedded proteins is referred to as plasma membrane.
Plasma	In physics and chemistry, a plasma is an ionized gas, and is usually considered to be a distinct phase of matter. "Ionized" in this case means that at least one electron has been dissociated from a significant fraction of the molecules.
Receptor potential	The change in membrane potential that results from sensory transduction is referred to as receptor potential.
Invertebrate	Invertebrate is a term coined by Jean-Baptiste Lamarck to describe any animal without a spinal column. It therefore includes all animals except vertebrates (fish, reptiles, amphibians, birds and mammals).
Skeletal muscle	Skeletal muscle is a type of striated muscle, attached to the skeleton. They are used to facilitate movement, by applying force to bones and joints; via contraction. They generally contract voluntarily (via nerve stimulation), although they can contract involuntarily.
Muscle	Muscle is a contractile form of tissue. It is one of the four major tissue types, the other three being epithelium, connective tissue and nervous tissue. Muscle contraction is used to move parts of the body, as well as to move substances within the body.
Stretch receptor	Stretch receptor refers to a type of mechanoreceptor sensitive to changes in muscle length; detects the position of body parts.

Tendon	A tendon or sinew is a tough band of fibrous connective tissue that connects muscle to bone. They are similar to ligaments except that ligaments join one bone to another.
Hair cell	The hair cell is a sensory cell of both the auditory system and the vestibular system in all vertebrates. In mammals, the auditory hair cells are located within the organ of Corti on a thin basilar membrane in the cochlea of the inner ear.
Organ	Organ refers to a structure consisting of several tissues adapted as a group to perform specific functions.
Eardrum	The tympanic membrane, colloquially known as the eardrum, is a thin membrane that separates the external ear from the middle ear. Its function is to transmit sound from the air to the ossicles inside the middle ear. The malleus bone bridges the gap between the eardrum and the other ossicles.
Tympanic membrane	Tympanic membrane is a thin membrane that separates the outer ear from the middle ear. Its function is to transmit sound from the air to the ossicles inside the middle ear.
Middle ear	Middle ear refers to one of three main regions of the vertebrate ear; a chamber containing three small bones that convey vibrations from the eardrum to the inner ear.
Eustachian tube	The Eustachian tube (or auditory tube) is a tube that links the pharynx to the middle ear. Normally the Eustachian tube is closed, but it can open to let a small amount of air through to equalize the pressure between the middle ear and the atmosphere.
Oval window	Oval window in the vertebrate ear is a membrane-covered gap in the skull bone, through which sound waves pass from the middle ear into the inner ear.
Inner ear	The Inner ear consists of the cochlea, where a wave is created by a difference in pressure between the scala vestibuli and the scala tympani. The cochlea has three fluid filled sections. The perilymph fluid in the canals differs from the endolymph fluid in the cochlear duct. The organ of Corti is the sensor of pressure variations.
Organ of corti	The organ of Corti is the organ in the inner ear of mammals that contains the auditory sensory cells, the so-called hair cells. It is situated on the basilar membrane and protrudes into the scala media. It contains four rows of hair cells whose hair bundles stick out from its surface.
Basilar membrane	The basilar membrane within the cochlea of the inner ear separates two liquid filled tubes that run along the coil of the cochlea, the scala media and the scala tympani. This separation is the main function of the basilar membrane in the hearing organ of all land vertebrates.
Stapes	The stapes is the stirrup-shaped small bone or ossicle in the middle ear which attaches the incus to the fenestra ovalis, the "oval window" which is adjacent to the vestibule of the inner ear. It is the smallest bone in the human body. The stapes transmits the sound vibrations from the incus to the membrane of the fenestra ovalis.
Round window	Round window refers to membrane-covered opening between the inner ear and the middle ear.
Conduction	Conduction refers to the direct transfer of thermal motion between molecules of objects in direct contact with each other. Also refers to the conduction of heat and charged particles through matter or the conduction of signals along nerve cells.
Nerve	A nerve is an enclosed, cable-like bundle of nerve fibers or axons, which includes the glia that ensheath the axons in myelin.
Rhodopsin	Rhodopsin is expressed in vertebrate photoreceptor cells. It is a pigment of the retina that is responsible for both the formation of the photoreceptor cells and the first events in the perception of light. Rhodopsins belong to the class of G-protein coupled receptors. It is the

chemical that allows night-vision, and is extremely sensitive to light.

Retinal	Retinal is fundamental in the transduction of light into visual signals in the photoreceptor level of the retina.
Retina	The retina is a thin layer of cells at the back of the eyeball of vertebrates and some cephalopods; it is the part of the eye which converts light into nervous signals.
Rod cell	Rod cell refers to photoreceptor cells in the retina of the eye that can function in less intense light than can the other type of photoreceptor, cone cells. Since they are more light-sensitive, they are responsible for night vision. There are about 100 million of them in the human retina.
Organelle	In cell biology, an organelle is one of several structures with specialized functions, suspended in the cytoplasm of a eukaryotic cell.
Synaptic terminal	Synaptic terminal is a bulb at the end of an axon in which neurotransmitter molecules are stored and released.
Compound eye	Compound eye refers to a type of eye, found in arthropods, that is composed of numerous independent subunits called ommatidia. Each ommatidium apparently contributes a piece of a mosaiclike image perceived by the animal.
Eye	An eye is an organ that detects light. Different kinds of light-sensitive organs are found in a variety of creatures.
Ommatidia	Ommatidia refers to the functional units of a compound eye; each includes a cornea, lens, and photoreceptor cells.
Sclera	The sclera is the (usually) white outer coating of the eye made of tough fibrin connective tissue which gives the eye its shape and helps to protect the delicate inner parts.
Cornea	The cornea is the transparent front part of the eye that covers the iris, pupil, and anterior chamber and provides most of an eye's optical power.
Iris	The colored part of the vertebrate eye, formed by the anterior portion of the choroid is called the iris.
Lens	The lens or crystalline lens is a transparent, biconvex structure in the eye that, along with the cornea, helps to refract light to focus on the retina. Its function is thus similar to a man-made optical lens.
Visual field	Area of vision for each eye is referred to as visual field.
Fovea	The fovea, a part of the eye, is a spot located in the center of the macula. The fovea is responsible for sharp central vision. At the center of the fovea there is a pit with a diameter of about 0.2 mm. It has a high concentration of cone cells and virtually no rods.
Color vision	Ability to detect the color of an object, dependent on three kinds of cone cells is called color vision.
Cone	Cone refers to a reproductive structure of gymnosperms that produces pollen in males or eggs in females.
Rods	Rods, are photoreceptor cells in the retina of the eye that can function in less intense light than can the other type of photoreceptor, cone cells.
Ganglion	In vertebrate anatomy, a ganglion is a tissue mass that contains the dendrites and cell bodies (or "somata") of nerve cells, in most case ones belonging to the peripheral nervous system.
Ganglion cell	A ganglion cell is a type of neuron located in the retina of the eye that receives visual

Go to **Cram101.com** for the Practice Tests for this Chapter.

information from photoreceptors via various intermediate cells such as bipolar cells, amacrine cells, and horizontal cells. The axons are myelinated.

Radiation

The emission of electromagnetic waves by all objects warmer than absolute zero is referred to as radiation.

Echolocation

Echolocation, also called Biosonar, is the biological sonar used by several mammals such as bats, dolphins and possibly whales. The term was coined by Donald Griffin, who was the first to conclusively demonstrate its existence in bats.

Go to **Cram101.com** for the Practice Tests for this Chapter.

Brain	The part of the central nervous system involved in regulating and controlling body activity and interpreting information from the senses transmitted through the nervous system is referred to as the brain.
Peripheral nervous system	The peripheral nervous system consists of the nerves and neurons that reside or extend outside the central nervous system--to serve the limbs and organs. The peripheral nervous system is divided into the somatic nervous system and the autonomic nervous system.
Nervous system	The nervous system of an animal coordinates the activity of the muscles, monitors the organs, constructs and processes input from the senses, and initiates actions.
Blood	Blood is a circulating tissue composed of fluid plasma and cells. The main function of blood is to supply nutrients (oxygen, glucose) and constitutional elements to tissues and to remove waste products.
Nerve	A nerve is an enclosed, cable-like bundle of nerve fibers or axons, which includes the glia that ensheath the axons in myelin.
Neural tube	The neural tube is the embryonal structure that gives rise to the brain and spinal cord. The neural tube is derived from a thickened area of ectoderm, the neural plate. The process of formation of the neural tube is called neurulation.
Hindbrain	Hindbrain refers to one of three ancestral and embryonic regions of the vertebrate brain; develops into the medulla oblongata, pons, and cerebellum.
Pons	The pons is a knob on the brain stem. It is part of the autonomic nervous system, and relays sensory information between the cerebellum and cerebrum. Some theories posit that it has a role in dreaming.
Spinal cord	The spinal cord is a part of the vertebrate nervous system that is enclosed in and protected by the vertebral column (it passes through the spinal canal). It consists of nerve cells. The spinal cord carries sensory signals and motor innervation to most of the skeletal muscles in the body.
Cerebellum	The cerebellum is a region of the brain that plays an important role in the integration of sensory perception and motor output. The cerebellum integrates these two functions, using the constant feedback on body position to fine-tune motor movements.
Midbrain	Midbrain refers to one of three ancestral and embryonic regions of the vertebrate brain; develops into sensory integrating and relay centers that send sensory information to the cerebrum.
Forebrain	Forebrain refers to one of three ancestral and embryonic regions of the vertebrate brain; develops into the thalamus, hypothalamus, and cerebrum.
Diencephalon	The diencephalon is the region of the brain that includes the epithalamus, thalamus, and hypothalamus. It is located above the mesencephalon of the brain stem. Sensory information is relayed between the brain stem and the rest of the brain regions.
Thalamus	The thalamus is located in the center of the brain, beneath the cerebral hemispheres and next to the third ventricle. It is formed of grey matter and can be thought of as a relay station for nerve impulses in the brain.
Medulla	Medulla in general means the inner part, and derives from the Latin word for 'marrow'. In medicine it is contrasted to the cortex.
Spinal nerve	The spinal nerve is usually a mized nerve, formed from the dorsal and ventral roots that come out of the spinal cord.
Ventral	The surface or side of the body normally oriented upwards, away from the pull of gravity, is the dorsal side; the opposite side, typically the one closest to the ground when walking on

all legs, swimming or flying, is the ventral side.

Flexor	A muscle or tendon which bends a limb, part of a limb, or part of the body is a flexor.
Muscle	Muscle is a contractile form of tissue. It is one of the four major tissue types, the other three being epithelium, connective tissue and nervous tissue. Muscle contraction is used to move parts of the body, as well as to move substances within the body.
Antagonistic muscles	Antagonistic muscles refers to a pair of muscles, one of which contracts and in so doing extends the other; an arrangement that makes possible movement of the skeleton at joints.
Dorsal	In anatomy, the dorsal is the side in which the backbone is located. This is usually the top of an animal, although in humans it refers to the back.
Brain stem	The brain stem refers to a composite substructure of the brain. It includes the midbrain, the pons and the medulla oblongata. It is the major route for communication between the forebrain, the spinal cord, and peripheral nerves. It also controls various functions including respiration, regulation of heart rhythms, and primary aspects of sound localization.
Stem	Stem refers to that part of a plant's shoot system that supports the leaves and reproductive structures.
Limbic system	The limbic system is a group of brain structures that are involved in various emotions such as aggression, fear, pleasure and also in the formation of memory. It affects the endocrine system and the autonomic nervous system. It consists of several subcortical structures located around the thalamus.
Cortex	In anatomy and zoology the cortex is the outermost or superficial layer of an organ or the outer portion of the stem or root of a plant.
Cerebral cortex	The cerebral cortex is a brain structure in vertebrates. It is the outermost layer of the cerebrum and has a grey color. In the "higher" animals, the surface becomes folded. The cerebral cortex, made up of four lobes, is involved in many complex brain functions including memory, attention, perceptual awareness, "thinking", language and consciousness.
Cerebral hemisphere	Cerebral hemisphere refers to the right or left half of the vertebrate cerebrum.
Stimulus	Stimulus in a nervous system, a factor that triggers sensory transduction.
Style	The style is a stalk connecting the stigma with the ovary below containing the transmitting tract, which facilitates the movement of the male gamete to the ovule.
Neuron	The neuron is a major class of cells in the nervous system. In vertebrates, they are found in the brain, the spinal cord and in the nerves and ganglia of the peripheral nervous system, and their primary role is to process and transmit neural information.
Homo sapiens	Homo sapiens are bipedal primates of the superfamily Hominoidea, together with the other apes—chimpanzees, gorillas, orangutans, and gibbons. They are the dominant sentient species on planet Earth.
Homo sapien	Humans, or human beings, are classified as bipedal primates belonging to the mammalian species Homo sapien. Humans have a highly developed brain capable of abstract reasoning, language, and introspection.
Autonomic nervous system	The autonomic nervous system is the part of the nervous system that is not consciously controlled. It is commonly divided into two usually antagonistic subsystems: the sympathetic and parasympathetic nervous system.
Parasympathetic	Parasympathetic division refers to one of two sets of neurons in the autonomic nervous

Go to **Cram101.com** for the Practice Tests for this Chapter.

division	system. It generally promotes body activities that gain and conserve energy, such as digestion and reduced heart rate.
Sympathetic division	Sympathetic division refers to one of two sets of neurons in the autonomic nervous system. It generally prepares the body for energy-consuming activities, such as fleeing or fighting. It also is a subdivision of the autonomic nervous system.
Sympathetic	The sympathetic nervous system activates what is often termed the "fight or flight response". It is an automatic regulation system, that is, one that operates without the intervention of conscious thought.
Photoreceptor	A photoreceptor is a specialized type of neuron that is capable of phototransduction. More specifically, the photoreceptor sends signals to other neurons by a change in its membrane potential when it absorbs photons.
Axon	An axon is a long slender projection of a nerve cell, or neuron, which conducts electrical impulses away from the neuron's cell body or soma. They are in effect the primary transmission lines of the nervous system, and as bundles they help make up nerves.
Retina	The retina is a thin layer of cells at the back of the eyeball of vertebrates and some cephalopods; it is the part of the eye which converts light into nervous signals.
Retinal	Retinal is fundamental in the transduction of light into visual signals in the photoreceptor level of the retina.
Vertebrate	Vertebrate is a subphylum of chordates, specifically, those with backbones or spinal columns. They started to evolve about 530 million years ago during the Cambrian explosion, which is part of the Cambrian period.
Electroencep-alogram	Electroencephalogram refers to a graph that is the neurophysiologic measurement of the electrical activity of the brain by recording from electrodes placed on the scalp or, in special cases, on the cortex.
Resting potential	The resting potential of a cell is the membrane potential that would be maintained if there were no action potentials, synaptic potentials, or other active changes in the membrane potential.
Plasma membrane	Membrane surrounding the cytoplasm that consists of a phospholipid bilayer with embedded proteins is referred to as plasma membrane.
Plasma	In physics and chemistry, a plasma is an ionized gas, and is usually considered to be a distinct phase of matter. "Ionized" in this case means that at least one electron has been dissociated from a significant fraction of the molecules.
Threshold	Electrical potential level at which an action potential or nerve impulse is produced is called threshold.
Long-term depression	Long-term depression is a weakening of a synapse that lasts from hours to days. It results from either strong synaptic stimulation (as occurs in the cerebellum Purkinje cells) to persistent weak synaptic stimulation (as in the hippocampus).

Muscle	Muscle is a contractile form of tissue. It is one of the four major tissue types, the other three being epithelium, connective tissue and nervous tissue. Muscle contraction is used to move parts of the body, as well as to move substances within the body.
Adaptation	A biological adaptation is an anatomical structure, physiological process or behavioral trait of an organism that has evolved over a period of time by the process of natural selection such that it increases the expected long-term reproductive success of the organism.
Microtubule	Microtubule is a protein structure found within cells, one of the components of the cytoskeleton. They have diameter of ~ 24 nm and varying length from several micrometers to possible millimeters in axons of nerve cells. They serve as structural components within cells and are involved in many cellular processes including mitosis, cytokinesis, and vesicular transport.
Cilia	Numerous short, hairlike structures projecting from the cell surface that enable locomotion are cilia.
Cell	The cell is the structural and functional unit of all living organisms, and is sometimes called the "building block of life."
Invertebrate	Invertebrate is a term coined by Jean-Baptiste Lamarck to describe any animal without a spinal column. It therefore includes all animals except vertebrates (fish, reptiles, amphibians, birds and mammals).
Cilium	A cilium is an organelle projecting from a eukaryotic cell. They are extensions of the plasma membrane containing doublets of parallel microtubules.
Stroke	A stroke or cerebrovascular accident (CVA) occurs when the blood supply to a part of the brain is suddenly interrupted.
Flagella	Flagella are whip-like organelle that many unicellular organisms, and some multicellular ones, use to move about.
Sperm	Sperm refers to the male sex cell with three distinct parts at maturity: head, middle piece, and tail.
Protein	A protein is a complex, high-molecular-weight organic compound that consists of amino acids joined by peptide bonds. They are essential to the structure and function of all living cells and viruses. Many are enzymes or subunits of enzymes.
Cytoskeleton	Cytoskeleton refers to a meshwork of fine fibers in the cytoplasm of a eukaryotic cell; includes microfilaments, intermediate filaments, and microtubules.
Microfilament	Microfilament refers to the thinnest of the three main kinds of protein fibers making up the cytoskeleton of a eukaryotic cell; a solid, helical rod composed of the globular protein actin.
Actin	Actin is a globular protein that polymerizes helically forming filaments, which like the other two components of the cellular cytoskeleton form a three-dimensional network inside an eukaryotic cell. They provide mechanical support for the cell, determine the cell shape, enable cell movements .
Multicellular	Multicellular organisms are those organisms consisting of more than one cell, and having differentiated cells that perform specialized functions. Most life that can be seen with the naked eye is multicellular, as are all animals (i.e. members of the kingdom Animalia) and plants (i.e. members of the kingdom Plantae).
Protist	A protist is a heterogeneous group of living things, comprising those eukaryotes that are neither animals, plants, nor fungi. They are a paraphyletic grade, rather than a natural group, and do not have much in common besides a relatively simple organization

Cytoplasm	Cytoplasm refers to everything inside a cell between the plasma membrane and the nucleus; consists of a semifluid medium and organelles.
Pseudopod	Pseudopod is a temporary projection of a eukaryotic cell. Cells having this faculty are generally referred to as amoeboids.
Smooth muscle	Smooth muscle is a type of non-striated muscle, found within the "walls" of hollow organs; such as blood vessels, the bladder, the uterus, and the gastrointestinal tract. Smooth muscle is used to move matter within the body, via contraction; it generally operates "involuntarily", without nerve stimulation.
Action potential	An action potential is a wave of electrical discharge that travels along the membrane of a cell. They communicate fast internal messages between tissues making them an essential feature of animal life at the microscopic level.
Cardiac muscle	Cardiac muscle is a type of striated muscle found within the heart. Its function is to "pump" blood through the circulatory system. Unlike skeletal muscle, which contracts in response to nerve stimulation, and like smooth muscle, cardiac muscle is myogenic, meaning that it stimulates its own contraction without a requisite electrical impulse.
Ion	Ion refers to an atom or molecule that has gained or lost one or more electrons, thus acquiring an electrical charge.
Depolarization	Depolarization is a decrease in the absolute value of a cell's membrane potential.
Autonomic nervous system	The autonomic nervous system is the part of the nervous system that is not consciously controlled. It is commonly divided into two usually antagonistic subsystems: the sympathetic and parasympathetic nervous system.
Nervous system	The nervous system of an animal coordinates the activity of the muscles, monitors the organs, constructs and processes input from the senses, and initiates actions.
Muscle fiber	Cell with myofibrils containing actin and myosin filaments arranged within sarcomeres is a muscle fiber.
Fiber	Fiber is a class of materials that are continuous filaments or are in discrete elongated pieces, similar to lengths of thread. They are of great importance in the biology of both plants and animals, for holding tissues together.
Myofibril	Myofibril is a cylindrical organelle, found within muscle cells. They are bundles of filaments that run from one end of the cell to the other and are attached to the cell surface membrane at each end.
Sarcomere	A sarcomere is the basic unit of a cross striated muscle's myofibril. They are multi-protein complexes composed of three different filament systems. A sarcomere is defined as the segment between two neighboring Z-lines (or Z-discs).
Myosin	Myosin is a large family of motor proteins found in eukaryotic tissues. They are responsible for actin-based motility.
Motor neuron	A nerve cell that conveys command signals from the central nervous system to effector cells, such as muscle cells or gland cells is a motor neuron.
Neuron	The neuron is a major class of cells in the nervous system. In vertebrates, they are found in the brain, the spinal cord and in the nerves and ganglia of the peripheral nervous system, and their primary role is to process and transmit neural information.
Synapse	A junction, or relay point, between two neurons, or between a neuron and an effector cell. Electrical and chemical signals are relayed from one cell to another at a synapse.
Threshold	Electrical potential level at which an action potential or nerve impulse is produced is

Go to **Cram101.com** for the Practice Tests for this Chapter.

called threshold.

Plasma membrane	Membrane surrounding the cytoplasm that consists of a phospholipid bilayer with embedded proteins is referred to as plasma membrane.
Plasma	In physics and chemistry, a plasma is an ionized gas, and is usually considered to be a distinct phase of matter. "Ionized" in this case means that at least one electron has been dissociated from a significant fraction of the molecules.
Sarcoplasmic reticulum	Smooth endoplasmic reticulum of skeletal muscle cells specially adapted for calcium ion storage and release is called sarcoplasmic reticulum.
Filament	The stamen is the male organ of a flower. Each stamen generally has a stalk called the filament, and, on top of the filament, an anther. The filament is a long chain of proteins, such as those found in hair, muscle, or in flagella.
Troponin	A molecule found in thin filaments of muscle that helps regulate when muscle cells contract is referred to as troponin.
T tubule	A T tubule is a deep invagination of the plasma membrane found in skeletal and cardiac muscle cells. These invaginations allow depolarization of the membrane to quickly penetrate to the interior of the cell.
Enzyme	An enzyme is a protein that catalyzes, or speeds up, a chemical reaction. They are essential to sustain life because most chemical reactions in biological cells would occur too slowly, or would lead to different products, without them.
Skeletal muscle	Skeletal muscle is a type of striated muscle, attached to the skeleton. They are used to facilitate movement, by applying force to bones and joints; via contraction. They generally contract voluntarily (via nerve stimulation), although they can contract involuntarily.
Runner	A horizontally growing stem that may develop new plants at nodes that touch the soil is called runner.
Skeleton	In biology, the skeleton or skeletal system is the biological system providing physical support in living organisms.
Hydrostatic skeleton	A hydrostatic skeleton or hydroskeleton is a structure found in many soft-bodied invertebrates consisting of a fluid-filled cavity surrounded by muscles. The pressure of the fluid and action of the surrounding muscles are used to change an organism's shape and produce movement.
Body cavity	A fluid-containing space between the digestive tract and the body wall is referred to as body cavity.
Substrate	A substrate is a molecule which is acted upon by an enzyme. Each enzyme recognizes only the specific substrate of the reaction it catalyzes. A surface in or on which an organism lives.
Exoskeleton	An exoskeleton is an external anatomical feature that supports and protects an animal's body. Many invertebrate animals such as insects, crustaceans and shellfish have an exoskeleton.
Cuticle	Cuticle in animals, a tough, nonliving outer layer of the skin. In plants, a waxy coating on the surface of stems and leaves that helps retain water.
Arthropod	Arthropod refers to member of a phylum of invertebrates that contains among other groups crustaceans and insects that have an exoskeleton and jointed appendages.
Salt	Salt is a term used for ionic compounds composed of positively charged cations and negatively charged anions, so that the product is neutral and without a net charge.
Molting	In animals, molting is the routine shedding off old feathers in birds, or of old skin in reptiles, or of old hairs in mammals. In arthropods, such as insects, arachnids and

crustaceans, molting describes the shedding of its exoskeleton (which is often called its shell), typically to let it grow.

Endoskeleton	An endoskeleton is an internal support structure of an animal. Three phyla of animals possess endoskeletons of various complexity: Chordata, Echinodermata, and Porifera. the endoskeleton allows the body to move and gives the body structure and shape.
Matrix	In biology, matrix (plural: matrices) is the material between animal or plant cells, the material (or tissue) in which more specialized structures are embedded, and a specific part of the mitochondrion that is the site of oxidation of organic molecules.
Tissue	Group of similar cells which perform a common function is called tissue.
Cartilaginous fish	A fish that has a flexible skeleton made of cartilage is a cartilaginous fish.
Vertebrate	Vertebrate is a subphylum of chordates, specifically, those with backbones or spinal columns. They started to evolve about 530 million years ago during the Cambrian explosion, which is part of the Cambrian period.
Cartilage	Cartilage is a type of dense connective tissue. Cartilage is composed of cells called chondrocytes which are dispersed in a firm gel-like ground substance, called the matrix. Cartilage is avascular (contains no blood vessels) and nutrients are diffused through the matrix.
Compact bone	Type of bone that contains osteons consisting of concentric layers of matrix and osteocytes in lacunae is called compact bone. It forms the stout walls of the diaphysis of long bones and a thin wall of the epiphysis of long bones
Connective tissue	Connective tissue is any type of biological tissue with an extensive extracellular matrix and often serves to support, bind together, and protect organs.
Tendon	A tendon or sinew is a tough band of fibrous connective tissue that connects muscle to bone. They are similar to ligaments except that ligaments join one bone to another.
Element	A chemical element, often called simply element, is a chemical substance that cannot be divided or changed into other chemical substances by any ordinary chemical technique. An element is a class of substances that contain the same number of protons in all its atoms.
Blood	Blood is a circulating tissue composed of fluid plasma and cells. The main function of blood is to supply nutrients (oxygen, glucose) and constitutional elements to tissues and to remove waste products.
Species	Group of similarly constructed organisms capable of interbreeding and producing fertile offspring is a species.
Nematocyst	In cnidarians, a capsule that contains a threadlike fiber whose release aids in the capture of prey is referred to as nematocyst.
Pigment	Pigment is any material resulting in color in plant or animal cells which is the result of selective absorption.
Effector	An effector is a molecule (originally referring to small molecules but now encompassing any regulatory molecule, includes proteins) that binds to a protein and thereby alters the activity of that protein.
Skin	Skin is an organ of the integumentary system composed of a layer of tissues that protect underlying muscles and organs.
Acetylcholine	The chemical compound acetylcholine was the first neurotransmitter to be identified. It is a chemical transmitter in both the peripheral nervous system (PNS) and central nervous system

Go to **Cram101.com** for the Practice Tests for this Chapter.

(CNS) in many organisms including humans.

Voltage

A measure of the electrical difference that exists between two different points or objects is called voltage.

Organ

Organ refers to a structure consisting of several tissues adapted as a group to perform specific functions.

Metabolism	Metabolism is the biochemical modification of chemical compounds in living organisms and cells. This includes the biosynthesis of complex organic molecules (anabolism) and their breakdown (catabolism).
Diffusion	Diffusion refers to the spontaneous movement of particles of any kind from where they are more concentrated to where they are less concentrated.
Cellular respiration	Cellular respiration is the process in which the chemical bonds of energy-rich molecules such as glucose are converted into energy usable for life processes.
Respiration	Respiration is the process by which an organism obtains energy by reacting oxygen with glucose to give water, carbon dioxide and ATP (energy). Respiration takes place on a cellular level in the mitochondria of the cells and provide the cells with energy.
Invertebrate	Invertebrate is a term coined by Jean-Baptiste Lamarck to describe any animal without a spinal column. It therefore includes all animals except vertebrates (fish, reptiles, amphibians, birds and mammals).
Ectothermic	Ectothermic creatures control their body temperature through external means, such as the sun, or flowing air/water.
Partial pressure	The partial pressure of a gas in a mixture or solution is what the pressure of that gas would be if all other components of the mixture or solution suddenly vanished without its temperature changing.
Gradient	Gradient refers to a difference in concentration, pressure, or electrical charge between two regions.
Insect	An arthropod that usually has three body segments , three pairs of legs, and one or two pairs of wings is called an insect. They are the largest and (on land) most widely-distributed taxon within the phylum Arthropoda. They comprise the most diverse group of animals on the earth, with around 925,000 species described
Cell	The cell is the structural and functional unit of all living organisms, and is sometimes called the "building block of life."
Gill	An extension of the body surface of an aquatic animal, specialized for gas exchange and/or suspension feeding is called a gill.
Blood	Blood is a circulating tissue composed of fluid plasma and cells. The main function of blood is to supply nutrients (oxygen, glucose) and constitutional elements to tissues and to remove waste products.
Species	Group of similarly constructed organisms capable of interbreeding and producing fertile offspring is a species.
Mammal	Homeothermic vertebrate characterized especially by the presence of hair and mammary glands is a mammal.
Evolution	In biology, evolution is the process by which novel traits arise in populations and are passed on from generation to generation. Its action over large stretches of time explains the origin of new species and ultimately the vast diversity of the biological world.
Ventilation	Ventilation refers to a mechanism that provides contact between an animal's respiratory surface and the air or water to which it is exposed. It is also called breathing.
Residual volume	Residual volume is the amount of air left in the lungs after a maximal exhalation. This averages about 1.5 L..
Pharynx	The pharynx is the part of the digestive system and respiratory system of many animals immediately behind the mouth and in front of the esophagus.

Trachea	Trachea is an airway through which respiratory gas transport takes place in organisms. In terrestrial vertebrates, such as birds and humans, the trachea lets air move from the throat to the lungs. In terrestrial invertebrates, such as onychophorans and beetles, they conduct air from outside the organism directly to all of its internal tissues.
Cartilage	Cartilage is a type of dense connective tissue. Cartilage is composed of cells called chondrocytes which are dispersed in a firm gel-like ground substance, called the matrix. Cartilage is avascular (contains no blood vessels) and nutrients are diffused through the matrix.
Bronchiole	The bronchiole is the first airway branch that no longer contains cartilage. They are branches of the bronchi, and are smaller than one millimetre in diameter.
Alveoli	Alveoli are anatomical structures that have the form of a hollow cavity. In the lung, the pulmonary alveoli are spherical outcroppings of the respiratory bronchioles and are the primary sites of gas exchange with the blood.
Capillary	A capillary is the smallest of a body's blood vessels, measuring 5-10 micro meters. They connect arteries and veins, and most closely interact with tissues. Their walls are composed of a single layer of cells, the endothelium. This layer is so thin that molecules such as oxygen, water and lipids can pass through them by diffusion and enter the tissues.
Red blood cell	The red blood cell is the most common type of blood cell and is the vertebrate body's principal means of delivering oxygen from the lungs or gills to body tissues via the blood.
Surface tension	A measure of how difficult it is to stretch or break the surface of a liquid is referred to as surface tension.
Cohesion	The tendency of the molecules of a substance to stick together is referred to as cohesion.
Microorganism	A microorganism is an organism that is so small that it is microscopic (invisible to the naked eye). They are often illustrated using single-celled, or unicellular organisms; however, some unicellular protists are visible to the naked eye, and some multicellular species are microscopic.
Cilia	Numerous short, hairlike structures projecting from the cell surface that enable locomotion are cilia.
Diaphragm	The diaphragm is a shelf of muscle extending across the bottom of the ribcage. It is critically important in respiration: in order to draw air into the lungs, the diaphragm contracts, thus enlarging the thoracic cavity and reducing intra-thoracic pressure.
Circulatory system	The circulatory system or cardiovascular system is the organ system which circulates blood around the body of most animals.
Plasma	In physics and chemistry, a plasma is an ionized gas, and is usually considered to be a distinct phase of matter. "Ionized" in this case means that at least one electron has been dissociated from a significant fraction of the molecules.
Protein	A protein is a complex, high-molecular-weight organic compound that consists of amino acids joined by peptide bonds. They are essential to the structure and function of all living cells and viruses. Many are enzymes or subunits of enzymes.
Hemoglobin	Hemoglobin is the iron-containing oxygen-transport metalloprotein in the red cells of the blood in mammals and other animals. Hemoglobin transports oxygen from the lungs to the rest of the body, such as to the muscles, where it releases the oxygen load.
Molecule	A molecule is the smallest particle of a pure chemical substance that still retains its chemical composition and properties.
Brain	The part of the central nervous system involved in regulating and controlling body activity

and interpreting information from the senses transmitted through the nervous system is referred to as the brain.

Tissue	Group of similar cells which perform a common function is called tissue.
Polypeptide	Polypeptide refers to polymer of many amino acids linked by peptide bonds.
Muscle	Muscle is a contractile form of tissue. It is one of the four major tissue types, the other three being epithelium, connective tissue and nervous tissue. Muscle contraction is used to move parts of the body, as well as to move substances within the body.
Myoglobin	Myoglobin is a single-chain protein of 153 amino acids, containing a heme (iron-containing porphyrin) group in the center. With a molecular weight of 16,700 Daltons, it is the primary oxygen-carrying pigment of muscle tissues.
Fetus	Fetus refers to a developing human from the ninth week of gestation until birth; has all the major structures of an adult.
Bicarbonate ion	The bicarbonate ion consists of one central carbon atom surrounded by three identical oxygen atoms in a trigonal planar arrangement, with a hydrogen atom attached to one of the oxygens.
Ion	Ion refers to an atom or molecule that has gained or lost one or more electrons, thus acquiring an electrical charge.
Acid	An acid is a water-soluble, sour-tasting chemical compound that when dissolved in water, gives a solution with a pH of less than 7.
Carbonic anhydrase	Carbonic anhydrase is a family of zinc-containing enzymes that catalyze the rapid interconversion of carbon dioxide and water into carbonic acid, protons, and bicarbonate ions.
Reactant	A reactant is any substance initially present in a chemical reaction. These reactants react with each other to form the products of a chemical reaction. In a chemical equation, the reactants are the elements or compounds on the left hand side of the reaction equation.
Nervous system	The nervous system of an animal coordinates the activity of the muscles, monitors the organs, constructs and processes input from the senses, and initiates actions.
Autonomic nervous system	The autonomic nervous system is the part of the nervous system that is not consciously controlled. It is commonly divided into two usually antagonistic subsystems: the sympathetic and parasympathetic nervous system.
Spinal cord	The spinal cord is a part of the vertebrate nervous system that is enclosed in and protected by the vertebral column (it passes through the spinal canal). It consists of nerve cells. The spinal cord carries sensory signals and motor innervation to most of the skeletal muscles in the body.
Brain stem	The brain stem refers to a composite substructure of the brain. It includes the midbrain, the pons and the medulla oblongata. It is the major route for communication between the forebrain, the spinal cord, and peripheral nerves. It also controls various functions including respiration, regulation of heart rhythms, and primary aspects of sound localization.
Stem	Stem refers to that part of a plant's shoot system that supports the leaves and reproductive structures.
Medulla	Medulla in general means the inner part, and derives from the Latin word for 'marrow'. In medicine it is contrasted to the cortex.
Ventral	The surface or side of the body normally oriented upwards, away from the pull of gravity, is the dorsal side; the opposite side, typically the one closest to the ground when walking on

Go to **Cram101.com** for the Practice Tests for this Chapter.

all legs, swimming or flying, is the ventral side.

Blood	Blood is a circulating tissue composed of fluid plasma and cells. The main function of blood is to supply nutrients (oxygen, glucose) and constitutional elements to tissues and to remove waste products.
Circulatory system	The circulatory system or cardiovascular system is the organ system which circulates blood around the body of most animals.
Invertebrate	Invertebrate is a term coined by Jean-Baptiste Lamarck to describe any animal without a spinal column. It therefore includes all animals except vertebrates (fish, reptiles, amphibians, birds and mammals).
Tissue	Group of similar cells which perform a common function is called tissue.
Closed circulatory system	A closed circulatory system is found in all vertebrates, as well as of annelids (for example, earthworms) and cephalopods (squids and octopuses). The blood never leaves the system of blood vessels consisting of arteries, capillaries and veins.
Blood vessel	A blood vessel is a part of the circulatory system and function to transport blood throughout the body. The most important types, arteries and veins, are so termed because they carry blood away from or towards the heart, respectively.
Dorsal	In anatomy, the dorsal is the side in which the backbone is located. This is usually the top of an animal, although in humans it refers to the back.
Vascular	In botany vascular refers to tissues that contain vessels for transporting liquids. In anatomy and physiology, vascular means related to blood vessels, which are part of the Circulatory system.
Arteriole	An arteriole is a blood vessel that extends and branches out from an artery and leads to capillaries. They have thick muscular walls and are the primary site of vascular resistance.
Atrium	The atrium is the blood collection chamber of a heart. It has a thin-walled structure that allows blood to return to the heart. There is at least one atrium in an animal with a closed circulatory system
Ventricle	In the heart, a ventricle is a heart chamber which collects blood from an atrium (another heart chamber) and pumps it out of the heart.
Aorta	The largest artery in the human body, the aorta originates from the left ventricle of the heart and brings oxygenated blood to all parts of the body in the systemic circulation.
Gill	An extension of the body surface of an aquatic animal, specialized for gas exchange and/or suspension feeding is called a gill.
Systemic circulation	Systemic circulation is a circuit of circulation in the cardiovascular system. Blood circulates from the left ventricle to the organs and tissues to the systemic veins to the right atrium.
Amphibian	Amphibian is a taxon of animals that include all tetrapods (four-legged vertebrates) that do not have amniotic eggs.
Species	Group of similarly constructed organisms capable of interbreeding and producing fertile offspring is a species.
Pulmonary circuit	Pulmonary circuit refers to one of two main blood circuits in terrestrial vertebrates; conveys blood between the heart and the lungs.
Reptile	Member of a class of terrestrial vertebrates with internal fertilization, scaly skin, and an egg with a leathery shell is called reptile.
Vein	Vein in animals, is a vessel that returns blood to the heart. In plants, a vascular bundle in a leaf, composed of xylem and phloem.

Go to **Cram101.com** for the Practice Tests for this Chapter.

Diastole	The stage of the heart cycle in which the heart muscle is relaxed, allowing the chambers to fill with blood is called diastole.
Cardiac cycle	Cardiac cycle is the term used to describe the sequence of events that occur as a heart works to pump blood through the body. Every single 'beat' of the heart involves three major stages: atrial systole, ventricular systole and complete cardiac diastole.
Blood pressure	Blood pressure is the pressure exerted by the blood on the walls of the blood vessels.
Systole	The contraction stage of the heart cycle, when the heart chambers actively pump blood is systole.
Cardiac muscle	Cardiac muscle is a type of striated muscle found within the heart. Its function is to "pump" blood through the circulatory system. Unlike skeletal muscle, which contracts in response to nerve stimulation, and like smooth muscle, cardiac muscle is myogenic, meaning that it stimulates its own contraction without a requisite electrical impulse.
Muscle	Muscle is a contractile form of tissue. It is one of the four major tissue types, the other three being epithelium, connective tissue and nervous tissue. Muscle contraction is used to move parts of the body, as well as to move substances within the body.
Cell	The cell is the structural and functional unit of all living organisms, and is sometimes called the "building block of life."
Action potential	An action potential is a wave of electrical discharge that travels along the membrane of a cell. They communicate fast internal messages between tissues making them an essential feature of animal life at the microscopic level.
Depolarization	Depolarization is a decrease in the absolute value of a cell's membrane potential.
Nervous system	The nervous system of an animal coordinates the activity of the muscles, monitors the organs, constructs and processes input from the senses, and initiates actions.
Nerve	A nerve is an enclosed, cable-like bundle of nerve fibers or axons, which includes the glia that ensheath the axons in myelin.
Atrioventric-lar node	The atrioventricular node is the tissue between the atria and the ventricles of the heart, which conducts the normal electrical impulse from the atria to the ventricles.
Node	In botany, a node is the place on a stem where a leaf is attached.
Smooth muscle	Smooth muscle is a type of non-striated muscle, found within the "walls" of hollow organs; such as blood vessels, the bladder, the uterus, and the gastrointestinal tract. Smooth muscle is used to move matter within the body, via contraction; it generally operates "involuntarily", without nerve stimulation.
Brain	The part of the central nervous system involved in regulating and controlling body activity and interpreting information from the senses transmitted through the nervous system is referred to as the brain.
Capillary bed	Capillary bed refers to a network of capillaries that infiltrate every organ and tissue in the body. A layer of tissue densely packed with capillaries.
Capillary	A capillary is the smallest of a body's blood vessels, measuring 5-10 micro meters. They connect arteries and veins, and most closely interact with tissues. Their walls are composed of a single layer of cells, the endothelium. This layer is so thin that molecules such as oxygen, water and lipids can pass through them by diffusion and enter the tissues.
Hormone	A hormone is a chemical messenger from one cell to another. All multicellular organisms produce hormones. The best known hormones are those produced by endocrine glands of vertebrate animals, but hormones are produced by nearly every organ system and tissue type in

Go to **Cram101.com** for the Practice Tests for this Chapter.

a human or animal body. Hormone molecules are secreted directly into the bloodstream, they move by circulation or diffusion to their target cells, which may be nearby cells in the same tissue or cells of a distant organ of the body.

Voltage	A measure of the electrical difference that exists between two different points or objects is called voltage.
Liver	The liver is an organ in vertebrates, including humans. It plays a major role in metabolism and has a number of functions in the body including drug detoxification, glycogen storage, and plasma protein synthesis. It also produces bile, which is important for digestion.
Protein	A protein is a complex, high-molecular-weight organic compound that consists of amino acids joined by peptide bonds. They are essential to the structure and function of all living cells and viruses. Many are enzymes or subunits of enzymes.
Plasma	In physics and chemistry, a plasma is an ionized gas, and is usually considered to be a distinct phase of matter. "Ionized" in this case means that at least one electron has been dissociated from a significant fraction of the molecules.
Lymphatic system	Lymph originates as blood plasma lost from the circulatory system, which leaks out into the surrounding tissues. The lymphatic system collects this fluid by diffusion into lymph capillaries, and returns it to the circulatory system.
Tissue fluid	Tissue fluid is a solution which bathes and surrounds the cells of multicellular animals and is also called the intercellular fluid.
Lymph	Lymph originates as blood plasma lost from the circulatory system, which leaks out into the surrounding tissues. The lymphatic system collects this fluid by diffusion into lymph capillaries, and returns it to the circulatory system.
Lymph node	A lymph node acts as a filter, with an internal honeycomb of connective tissue filled with lymphocytes that collect and destroy bacteria and viruses. When the body is fighting an infection, these lymphocytes multiply rapidly and produce a characteristic swelling of the lymph node.
Cancer	Cancer is a class of diseases or disorders characterized by uncontrolled division of cells and the ability of these cells to invade other tissues, either by direct growth into adjacent tissue through invasion or by implantation into distant sites by metastasis.
Connective tissue	Connective tissue is any type of biological tissue with an extensive extracellular matrix and often serves to support, bind together, and protect organs.
Atherosclerosis	Atherosclerosis is a disease affecting the arterial blood vessel. It is commonly referred to as a "hardening" or "furring" of the arteries. It is caused by the formation of multiple plaques within the arteries.
Artery	Vessel that takes blood away from the heart to the tissues and organs of the body is called an artery.
Stroke	A stroke or cerebrovascular accident (CVA) occurs when the blood supply to a part of the brain is suddenly interrupted.
Cardiovascular disease	Cardiovascular disease refers to a set of diseases of the heart and blood vessels.
Transcription factor	In molecular biology, a transcription factor is a protein that binds DNA at a specific promoter or enhancer region or site, where it regulates transcription.
Transcription	Transcription is the process through which a DNA sequence is enzymatically copied by an RNA polymerase to produce a complementary RNA. Or, in other words, the transfer of genetic information from DNA into RNA.

Go to **Cram101.com** for the Practice Tests for this Chapter.

343

Kidney	The kidney is a bean-shaped excretory organ in vertebrates. Part of the urinary system, the kidneys filter wastes (especially urea) from the blood and excrete them, along with water, as urine.
Hemoglobin	Hemoglobin is the iron-containing oxygen-transport metalloprotein in the red cells of the blood in mammals and other animals. Hemoglobin transports oxygen from the lungs to the rest of the body, such as to the muscles, where it releases the oxygen load.
Red blood cell	The red blood cell is the most common type of blood cell and is the vertebrate body's principal means of delivering oxygen from the lungs or gills to body tissues via the blood.
Platelet	Cell fragment that is necessary to blood clotting is a platelet. They are the blood cell fragments that are involved in the cellular mechanisms that lead to the formation of blood clots.
Hemophilia	Hemophilia is the name of any of several hereditary genetic illnesses that impair the body's ability to control bleeding. Genetic deficiencies cause lowered plasma clotting factor activity so as to compromise blood-clotting; when a blood vessel is injured, a scab will not form and the vessel can continue to bleed excessively for a very long period of time.
Ion	Ion refers to an atom or molecule that has gained or lost one or more electrons, thus acquiring an electrical charge.
Norepinephrine	Norepinephrine is a catecholamine and a phenethylamine with chemical formula $C_8H_{11}NO_3$. It is released from the adrenal glands as a hormone into the blood, but it is also a neurotransmitter in the nervous system where it is released from noradrenergic neurons during synaptic transmission.
Skeletal muscle	Skeletal muscle is a type of striated muscle, attached to the skeleton. They are used to facilitate movement, by applying force to bones and joints; via contraction. They generally contract voluntarily (via nerve stimulation), although they can contract involuntarily.
Posterior pituitary	The posterior pituitary gland comprises the posterior lobe of the pituitary gland and is part of the endocrine system. Despite its name, the posterior pituitary gland is not a gland, rather, it is largely a collection of axonal projections from the hypothalamus that terminate behind the anterior pituitary gland.
Stretch receptor	Stretch receptor refers to a type of mechanoreceptor sensitive to changes in muscle length; detects the position of body parts.
Receptor	A receptor is a protein on the cell membrane or within the cytoplasm or cell nucleus that binds to a specific molecule (a ligand), such as a neurotransmitter, hormone, or other substance, and initiates the cellular response to the ligand. Receptor, in immunology, the region of an antibody which shows recognition of an antigen.
Cardiovascular system	The circulatory system or cardiovascular system is the organ system which circulates blood around the body of most animals.
Adaptation	A biological adaptation is an anatomical structure, physiological process or behavioral trait of an organism that has evolved over a period of time by the process of natural selection such that it increases the expected long-term reproductive success of the organism.
Acid	An acid is a water-soluble, sour-tasting chemical compound that when dissolved in water, gives a solution with a pH of less than 7.
Metabolism	Metabolism is the biochemical modification of chemical compounds in living organisms and cells. This includes the biosynthesis of complex organic molecules (anabolism) and their breakdown (catabolism).
Anaerobic	An anaerobic organism is any organism that does not require oxygen for growth.

Go to **Cram101.com** for the Practice Tests for this Chapter.

Calorie	Calorie refers to the amount of energy that raises the temperature of 1 g of water by 1°C.
Basal metabolic rate	Basal metabolic rate, is the rate of metabolism that occurs when an individual is at rest in a warm environment and is in the post absorptive state, and has not eaten for at least 12 hours.
Metabolic rate	Energy expended by the body per unit time is called metabolic rate.
Liver	The liver is an organ in vertebrates, including humans. It plays a major role in metabolism and has a number of functions in the body including drug detoxification, glycogen storage, and plasma protein synthesis. It also produces bile, which is important for digestion.
Glycogen	Glycogen refers to a complex, extensively branched polysaccharide of many glucose monomers; serves as an energy-storage molecule in liver and muscle cells.
Atrophy	Atrophy is the partial or complete wasting away of a part of the body. Causes of atrophy include poor nourishment, poor circulation, loss of hormonal support, loss of nerve supply to the target organ, disuse or lack of exercise, or disease intrinsic to the tissue itself.
Population	Group of organisms of the same species occupying a certain area and sharing a common gene pool is referred to as population.
Species	Group of similarly constructed organisms capable of interbreeding and producing fertile offspring is a species.
Blood pressure	Blood pressure is the pressure exerted by the blood on the walls of the blood vessels.
Blood	Blood is a circulating tissue composed of fluid plasma and cells. The main function of blood is to supply nutrients (oxygen, glucose) and constitutional elements to tissues and to remove waste products.
Acetyl	The acetyl radical contains a methyl group single-bonded to a carbonyl. The carbon of the carbonyl has an lone electron available, with which it forms a chemical bond to the remainder of the molecule.
Essential amino acid	An essential amino acid for an organism is an amino acid that cannot be synthesized by the organism from other available resources, and therefore must be supplied as part of its diet.
Amino acid	An amino acid is any molecule that contains both amino and carboxylic acid functional groups. They are the basic structural building units of proteins. They form short polymer chains called peptides or polypeptides which in turn form structures called proteins.
Acid	An acid is a water-soluble, sour-tasting chemical compound that when dissolved in water, gives a solution with a pH of less than 7.
Protein synthesis	The process whereby the tRNA utilizes the mRNA as a guide to arrange the amino acids in their proper sequence according to the genetic information in the chemical code of DNA is referred to as protein synthesis.
Protein	A protein is a complex, high-molecular-weight organic compound that consists of amino acids joined by peptide bonds. They are essential to the structure and function of all living cells and viruses. Many are enzymes or subunits of enzymes.
Immune system	The immune system is the system of specialized cells and organs that protect an organism from outside biological influences. When the immune system is functioning properly, it protects the body against bacteria and viral infections, destroying cancer cells and foreign substances.
Element	A chemical element, often called simply element, is a chemical substance that cannot be divided or changed into other chemical substances by any ordinary chemical technique. An element is a class of substances that contain the same number of protons in all its atoms.

Go to **Cram101.com** for the Practice Tests for this Chapter.

347

Urine	Concentrated filtrate produced by the kidneys and excreted via the bladder is called urine.
Atom	An atom is the smallest possible particle of a chemical element that retains its chemical properties.
Vitamin	A Vitamin is an organic molecule required by a living organism in minute amounts for proper health. An organism deprived of all sources of a particular vitamin will eventually suffer from disease symptoms specific to that vitamin.
Fruit	A fruit is the ripened ovary—together with seeds—of a flowering plant. In many species, the fruit incorporates the ripened ovary and surrounding tissues.
Lipid	Lipid is one class of aliphatic hydrocarbon-containing organic compounds essential for the structure and function of living cells. They are characterized by being water-insoluble but soluble in nonpolar organic solvents.
Skin	Skin is an organ of the integumentary system composed of a layer of tissues that protect underlying muscles and organs.
Microorganism	A microorganism is an organism that is so small that it is microscopic (invisible to the naked eye). They are often illustrated using single-celled, or unicellular organisms; however, some unicellular protists are visible to the naked eye, and some multicellular species are microscopic.
Anemia	Anemia is a deficiency of red blood cells and/or hemoglobin. This results in a reduced ability of blood to transfer oxygen to the tissues, and this causes hypoxia; since all human cells depend on oxygen for survival, varying degrees of anemia can have a wide range of clinical consequences.
Stomach	The stomach is an organ in the alimentary canal used to digest food. It's primary function is not the absorption of nutrients from digested food; rather, the main job of the stomach is to break down large food molecules into smaller ones, so that they can be absorbed into the blood more easily.
Hormone	A hormone is a chemical messenger from one cell to another. All multicellular organisms produce hormones. The best known hormones are those produced by endocrine glands of vertebrate animals, but hormones are produced by nearly every organ system and tissue type in a human or animal body. Hormone molecules are secreted directly into the bloodstream, they move by circulation or diffusion to their target cells, which may be nearby cells in the same tissue or cells of a distant organ of the body.
Goiter	Goiter refers to an enlargement of the thyroid gland resulting from a dietary iodine deficiency.
Nutrition	Nutrition refers to collectively, the processes involved in taking in, assimilating, and utilizing nutrients.
Leech	The leech is a annelid comprising the subclass Hirudinea. There are freshwater, terrestrial and marine leeches. Like their near relatives, the Oligochaeta, they share the presence of a clitellum.Like earthworms, leeches are hermaphrodites.
Herbivore	A herbivore is an animal that is adapted to eat primarily plant matter
Invertebrate	Invertebrate is a term coined by Jean-Baptiste Lamarck to describe any animal without a spinal column. It therefore includes all animals except vertebrates (fish, reptiles, amphibians, birds and mammals).
Carnivore	An animal that eats a diet consisting solely of meat is referred to as a carnivore.
Echolocation	Echolocation, also called Biosonar, is the biological sonar used by several mammals such as bats, dolphins and possibly whales. The term was coined by Donald Griffin, who was the first

to conclusively demonstrate its existence in bats.

Root	In vascular plants, the root is that organ of a plant body that typically lies below the surface of the soil. However, this is not always the case, since a root can also be aerial (that is, growing above the ground) or aerating (that is, growing up above the ground or especially above water).
Enzyme	An enzyme is a protein that catalyzes, or speeds up, a chemical reaction. They are essential to sustain life because most chemical reactions in biological cells would occur too slowly, or would lead to different products, without them.
Gastrovascular cavity	Gastrovascular cavity refers to a digestive compartment with a single opening, the mouth; may function in circulation, body support, waste disposal, and gas exchange, as well as digestion.
Cytoplasm	Cytoplasm refers to everything inside a cell between the plasma membrane and the nucleus; consists of a semifluid medium and organelles.
Plasma	In physics and chemistry, a plasma is an ionized gas, and is usually considered to be a distinct phase of matter. "Ionized" in this case means that at least one electron has been dissociated from a significant fraction of the molecules.
Radula	A toothed, grasping organ found in many mollusks, used to scrape up or shred food is referred to as a radula.
Ion	Ion refers to an atom or molecule that has gained or lost one or more electrons, thus acquiring an electrical charge.
Rectum	The rectum is the final straight portion of the large intestine in some mammals, and the gut in others, terminating in the anus.
Bacteria	The domain that contains procaryotic cells with primarily diacyl glycerol diesters in their membranes and with bacterial rRNA. Bacteria also is a general term for organisms that are composed of procaryotic cells and are not multicellular.
Vertebrate	Vertebrate is a subphylum of chordates, specifically, those with backbones or spinal columns. They started to evolve about 530 million years ago during the Cambrian explosion, which is part of the Cambrian period.
Carbohydrate	Carbohydrate is a chemical compound that contains oxygen, hydrogen, and carbon atoms. They consist of monosaccharide sugars of varying chain lengths and that have the general chemical formula $C_n(H_2O)_n$ or are derivatives of such.
Digestion	Digestion refers to the mechanical and chemical breakdown of food into molecules small enough for the body to absorb; the second main stage of food processing, following ingestion.
Cell	The cell is the structural and functional unit of all living organisms, and is sometimes called the "building block of life."
Tissue	Group of similar cells which perform a common function is called tissue.
Nerve	A nerve is an enclosed, cable-like bundle of nerve fibers or axons, which includes the glia that ensheath the axons in myelin.
Smooth muscle	Smooth muscle is a type of non-striated muscle, found within the "walls" of hollow organs; such as blood vessels, the bladder, the uterus, and the gastrointestinal tract. Smooth muscle is used to move matter within the body, via contraction; it generally operates "involuntarily", without nerve stimulation.
Muscle	Muscle is a contractile form of tissue. It is one of the four major tissue types, the other three being epithelium, connective tissue and nervous tissue. Muscle contraction is used to

Go to **Cram101.com** for the Practice Tests for this Chapter.

move parts of the body, as well as to move substances within the body.

Epiglottis	The epiglottis is a thin, lid-like flap of cartilage tissue covered with a mucous membrane, attached to the root of the tongue, that guards the entrance of the glottis, the opening between the vocal cords.
Peristalsis	Peristalsis is the process of involuntary wave-like successive muscular contractions by which food is moved through the digestive tract. The large, hollow organs of the digestive system contains muscles that enable their walls to move.
Pyloric sphincter	Pyloric sphincter in the vertebrate digestive tract, a muscular ring that regulates the passage of food out of the stomach and into the small intestine.
Sphincter	Muscle that surrounds a tube and closes or opens the tube by contracting and relaxing is referred to as sphincter.
Glucose	Glucose, a simple monosaccharide sugar, is one of the most important carbohydrates and is used as a source of energy in animals and plants. Glucose is one of the main products of photosynthesis and starts respiration.
Pepsin	Pepsin is a digestive protease released by the chief cells in the stomach that functions to degrade food proteins into peptides. It was the first animal enzyme to be discovered.
Chyme	Chyme is the liquid substance found in the stomach before passing the pyloric valve and entering the duodenum. It consists of partially digested food, water, hydrochloric acid, and various digestive enzymes.
Small intestine	The small intestine is the part of the gastrointestinal tract between the stomach and the large intestine (colon). In humans over 5 years old it is about 7m long. It is divided into three structural parts: duodenum, jejunum and ileum.
Intestine	The intestine is the portion of the alimentary canal extending from the stomach to the anus and, in humans and mammals, consists of two segments, the small intestine and the large intestine. The intestine is the part of the body responsible for extracting nutrition from food.
Duodenum	The duodenum is a hollow jointed tube connecting the stomach to the jejunum. It is the first part of the small intestine. Two very important ducts open into the duodenum, namely the bile duct and the pancreatic duct. The duodenum is largely responsible for the breakdown of food in the small intestine.
Gallbladder	The gallbladder is a pear-shaped organ that stores bile until the body needs it for digestion. It is connected to the liver and the duodenum by the biliary tract.
Bile	Bile is a bitter, greenish-yellow alkaline fluid secreted by the liver of most vertebrates. In many species, it is stored in the gallbladder between meals and upon eating is discharged into the duodenum where it aids the process of digestion.
Pancreas	The pancreas is a retroperitoneal organ that serves two functions: exocrine - it produces pancreatic juice containing digestive enzymes, and endocrine - it produces several important hormones, namely insulin.
Disaccharide	A disaccharide is a sugar (a carbohydrate) composed of two monosaccharides. The two monosaccharides are bonded via a condensation reaction.
Lactose	Lactose is a disaccharide that makes up around 2-8% of the solids in milk. Lactose is a disaccharide consisting of two subunits, a galactose and a glucose linked together.
Diffusion	Diffusion refers to the spontaneous movement of particles of any kind from where they are more concentrated to where they are less concentrated.

Absorption	Absorption is a physical or chemical phenomenon or a process in which atoms, molecules, or ions enter some bulk phase - gas, liquid or solid material. In nutrition, amino acids are broken down through digestion, which begins in the stomach.
Fiber	Fiber is a class of materials that are continuous filaments or are in discrete elongated pieces, similar to lengths of thread. They are of great importance in the biology of both plants and animals, for holding tissues together.
Colon	The colon is the part of the intestine from the cecum to the rectum. Its primary purpose is to extract water from feces.
Antibiotic	Antibiotic refers to substance such as penicillin or streptomycin that is toxic to microorganisms. Usually a product of a particular microorvanism or plant.
Large intestine	In anatomy of the digestive system, the colon, also called the large intestine or large bowel, is the part of the intestine from the cecum ('caecum' in British English) to the rectum. Its primary purpose is to extract water from feces.
Appendix	In human anatomy, the vermiform appendix is a blind ended tube connected to the cecum. It develops embryologically from the cecum.
Organic compound	An organic compound is any member of a large class of chemical compounds whose molecules contain carbon, with the exception of carbides, carbonates, carbon oxides and gases containing carbon.
Anaerobic	An anaerobic organism is any organism that does not require oxygen for growth.
Greenhouse gas	Greenhouse gas refers to a gas, such as carbon dioxide or methane, that traps sunlight energy in a planet's atmosphere as heat; a gas that participates in the greenhouse effect.
Cecum	The cecum is a pouch connected to the large intestine between the ileum. It is separated from the ileum by the ileocecal valve, and is considered to be the beginning of the colon. Its primary function is to absorb water and salts from undigested food.
Secretin	Secretin is a peptide hormone produced in the S cells of the duodenum. It is secreted in response to low duodenal pH or the presence of fatty acids in the duodenum, and stimulates the secretion of bicarbonate from the liver, pancreas, and duodenal Brunner's glands.
Circulatory system	The circulatory system or cardiovascular system is the organ system which circulates blood around the body of most animals.
Nervous system	The nervous system of an animal coordinates the activity of the muscles, monitors the organs, constructs and processes input from the senses, and initiates actions.
Glucagon	A peptide hormone secreted by islet cells in the pancreas that raises the level of glucose in the blood is referred to as glucagon. Glucagon is a 29 amino acid polypeptide acting as an important hormone in carbohydrate metabolism.
Insulin	Insulin is a polypeptide hormone that regulates carbohydrate metabolism. Apart from being the primary effector in carbohydrate homeostasis, it also has a substantial effect on small vessel muscle tone, controls storage and release of fat (triglycerides) and cellular uptake of both amino acids and some electrolytes.
Hypothalamus	Located below the thalamus, the hypothalamus links the nervous system to the endocrine system by synthesizing and secreting neurohormones often called releasing hormones because they function by stimulating the secretion of hormones from the anterior pituitary gland.
Brain	The part of the central nervous system involved in regulating and controlling body activity and interpreting information from the senses transmitted through the nervous system is referred to as the brain.

Go to **Cram101.com** for the Practice Tests for this Chapter.

Gene	Gene refers to a discrete unit of hereditary information consisting of a specific nucleotide sequence in DNA . Most of the genes of a eukaryote are located in its chromosomal DNA; a few are carried by the DNA of mitochondria and chloroplasts.
Leptin	Leptin is a 16 kDa protein hormone that plays a key role in metabolism and regulation of adipose tissue. It is released by fat cells in amounts mirroring overall body fat stores. Thus, circulating leptin levels give the brain a reading of energy storage for the purposes of regulating appetite and metabolism.
Predator	A predator is an animal or other organism that hunts and kills other organisms for food in an act called predation.
Cancer	Cancer is a class of diseases or disorders characterized by uncontrolled division of cells and the ability of these cells to invade other tissues, either by direct growth into adjacent tissue through invasion or by implantation into distant sites by metastasis.

Tissue fluid	Tissue fluid is a solution which bathes and surrounds the cells of multicellular animals and is also called the intercellular fluid.
Tissue	Group of similar cells which perform a common function is called tissue.
Homeostasis	Homeostasis is the property of an open system, especially living organisms, to regulate its internal environment to maintain a stable, constant condition, by means of multiple dynamic equilibrium adjustments, controlled by interrelated regulation mechanisms.
Blood	Blood is a circulating tissue composed of fluid plasma and cells. The main function of blood is to supply nutrients (oxygen, glucose) and constitutional elements to tissues and to remove waste products.
Vein	Vein in animals, is a vessel that returns blood to the heart. In plants, a vascular bundle in a leaf, composed of xylem and phloem.
Protein	A protein is a complex, high-molecular-weight organic compound that consists of amino acids joined by peptide bonds. They are essential to the structure and function of all living cells and viruses. Many are enzymes or subunits of enzymes.
Physiology	The study of the function of cells, tissues, and organs is referred to as physiology.
Vertebrate	Vertebrate is a subphylum of chordates, specifically, those with backbones or spinal columns. They started to evolve about 530 million years ago during the Cambrian explosion, which is part of the Cambrian period.
Solute	Substance that is dissolved in a solvent, forming a solution is referred to as a solute.
Urine	Concentrated filtrate produced by the kidneys and excreted via the bladder is called urine.
Species	Group of similarly constructed organisms capable of interbreeding and producing fertile offspring is a species.
Active transport	Active transport is the mediated transport of biochemicals, and other atomic/molecular substances, across membranes. In this form of transport, molecules move against either an electrical or concentration gradient.
Salt	Salt is a term used for ionic compounds composed of positively charged cations and negatively charged anions, so that the product is neutral and without a net charge.
Osmosis	Osmosis is the diffusion of a solvent through a semipermeable membrane from a region of low solute concentration to a region of high solute concentration.
Herbivore	A herbivore is an animal that is adapted to eat primarily plant matter
Nucleic acid	A nucleic acid is a complex, high-molecular-weight biochemical macromolecule composed of nucleotide chains that convey genetic information. The most common are deoxyribonucleic acid (DNA) and ribonucleic acid (RNA). They are found in all living cells and viruses.
Acid	An acid is a water-soluble, sour-tasting chemical compound that when dissolved in water, gives a solution with a pH of less than 7.
Ammonia	Ammonia is a compound of nitrogen and hydrogen with the formula NH_3. At standard temperature and pressure ammonia is a gas. It is toxic and corrosive to some materials, and has a characteristic pungent odor.
Excretion	Excretion is the biological process by which an organism chemically separates waste products from its body. The waste products are then usually expelled from the body by elimination.
Invertebrate	Invertebrate is a term coined by Jean-Baptiste Lamarck to describe any animal without a spinal column. It therefore includes all animals except vertebrates (fish, reptiles, amphibians, birds and mammals).

Cartilaginous fish	A fish that has a flexible skeleton made of cartilage is a cartilaginous fish.
Uric acid	An insoluble precipitate of nitrogenous waste excreted by land snails, insects, birds, and some reptiles is called uric acid.
Flame cell	A flame cell is a specialized excretory cell found in the Platyhelminthes (except the tubellarian order Acoela), these are the simplest animals to have a dedicated excretory system. It functions like a kidney removing waste materials.
Cell	The cell is the structural and functional unit of all living organisms, and is sometimes called the "building block of life."
Closed circulatory system	A closed circulatory system is found in all vertebrates, as well as of annelids (for example, earthworms) and cephalopods (squids and octopuses). The blood never leaves the system of blood vessels consisting of arteries, capillaries and veins.
Circulatory system	The circulatory system or cardiovascular system is the organ system which circulates blood around the body of most animals.
Coelom	A fluid filled body cavity with a complete lining called peritoneum derived from mesoderm is called a coelom. Organs formed inside a coelom can freely move, grow, and develop independently of the body wall while fluid cushions and protects them from shocks.
Insect	An arthropod that usually has three body segments , three pairs of legs, and one or two pairs of wings is called an insect. They are the largest and (on land) most widely-distributed taxon within the phylum Arthropoda. They comprise the most diverse group of animals on the earth, with around 925,000 species described
Muscle	Muscle is a contractile form of tissue. It is one of the four major tissue types, the other three being epithelium, connective tissue and nervous tissue. Muscle contraction is used to move parts of the body, as well as to move substances within the body.
Rectum	The rectum is the final straight portion of the large intestine in some mammals, and the gut in others, terminating in the anus.
Malpighian tubule	The Malpighian tubule is the insects' main organ of excretion and osmoregulation, helping them to maintain water and electrolyte balance.
Capillary bed	Capillary bed refers to a network of capillaries that infiltrate every organ and tissue in the body. A layer of tissue densely packed with capillaries.
Capillary	A capillary is the smallest of a body's blood vessels, measuring 5-10 micro meters. They connect arteries and veins, and most closely interact with tissues. Their walls are composed of a single layer of cells, the endothelium. This layer is so thin that molecules such as oxygen, water and lipids can pass through them by diffusion and enter the tissues.
Glomerulus	A glomerulus is a capillary tuft surrounded by Bowman's capsule in nephrons of the vertebrate kidney. It receives its blood supply from an afferent arteriole of the renal circulation, and empties into an efferent arteriole.
Arteriole	An arteriole is a blood vessel that extends and branches out from an artery and leads to capillaries. They have thick muscular walls and are the primary site of vascular resistance.
Filtration	Filtration involved in passive transport is the movement of water and solute molecules across the cell membrane due to hydrostatic pressure by the cardiovascular system.
Amphibian	Amphibian is a taxon of animals that include all tetrapods (four-legged vertebrates) that do not have amniotic eggs.
Skin	Skin is an organ of the integumentary system composed of a layer of tissues that protect

Go to **Cram101.com** for the Practice Tests for this Chapter.

underlying muscles and organs.

Kidney	The kidney is a bean-shaped excretory organ in vertebrates. Part of the urinary system, the kidneys filter wastes (especially urea) from the blood and excrete them, along with water, as urine.
Ureter	A ureter is a duct that carries urine from the kidneys to the urinary bladder. They are muscular tubes that can propel urine along by the motions of peristalsis.
Sphincter	Muscle that surrounds a tube and closes or opens the tube by contracting and relaxing is referred to as sphincter.
Bladder	A hollow muscular storage organ for storing urine is a bladder.
Medulla	Medulla in general means the inner part, and derives from the Latin word for 'marrow'. In medicine it is contrasted to the cortex.
Renal artery	The renal artery normally arise off the abdominal aorta and supply the kidneys with blood. The arterial supply of the kidneys is variable and there may be one or more supplying each kidney.
Artery	Vessel that takes blood away from the heart to the tissues and organs of the body is called an artery.
Cortex	In anatomy and zoology the cortex is the outermost or superficial layer of an organ or the outer portion of the stem or root of a plant.
Loop of henle	The loop of Henle is a section of the nephron that leads from the proximal convoluted tubule to the distal convoluted tubule in the kidney. Its primary function uses a countercurrent mechanism in the medulla to reabsorb water and ions from the urine.
Nephron	A nephron is the basic structural and functional unit of the kidney. Its chief function is to regulate water and soluble substances by filtering the blood, reabsorbing what is needed and excreting the rest as urine. They eliminate wastes from the body, regulate blood volume and pressure, control levels of electrolytes and metabolites, and regulate blood pH.
Proximal convoluted tubule	Highly coiled region of a nephron near the glomerular capsule, where tubular reabsorption takes place is called the proximal convoluted tubule.
Concentration gradient	Gradual change in chemical concentration from one point to another is called concentration gradient.
Gradient	Gradient refers to a difference in concentration, pressure, or electrical charge between two regions.
Renal cortex	The renal cortex is the outer portion of the kidney between the renal capsule and the renal medulla. It forms a continuous smooth outer zone in the adult which extends down between the pyramids.
Desert	A desert is a landscape form or region that receives little precipitation - less than 250 mm (10 in) per year. It is a biome characterized by organisms adapted to sparse rainfall and rapid evaporation.
Brain	The part of the central nervous system involved in regulating and controlling body activity and interpreting information from the senses transmitted through the nervous system is referred to as the brain.
Stretch receptor	Stretch receptor refers to a type of mechanoreceptor sensitive to changes in muscle length; detects the position of body parts.
Receptor	A receptor is a protein on the cell membrane or within the cytoplasm or cell nucleus that

binds to a specific molecule (a ligand), such as a neurotransmitter, hormone, or other substance, and initiates the cellular response to the ligand. Receptor, in immunology, the region of an antibody which shows recognition of an antigen.

Gene

Gene refers to a discrete unit of hereditary information consisting of a specific nucleotide sequence in DNA . Most of the genes of a eukaryote are located in its chromosomal DNA; a few are carried by the DNA of mitochondria and chloroplasts.

Hypothalamus

Located below the thalamus, the hypothalamus links the nervous system to the endocrine system by synthesizing and secreting neurohormones often called releasing hormones because they function by stimulating the secretion of hormones from the anterior pituitary gland.

Blood pressure

Blood pressure is the pressure exerted by the blood on the walls of the blood vessels.

Ion

Ion refers to an atom or molecule that has gained or lost one or more electrons, thus acquiring an electrical charge.

Buffer

A chemical substance that resists changes in pH by accepting H^+ ions from or donating H^+ ions to solutions is called a buffer.

Go to **Cram101.com** for the Practice Tests for this Chapter.

Species	Group of similarly constructed organisms capable of interbreeding and producing fertile offspring is a species.
Hybridization	In molecular biology hybridization is the process of joining two complementary strands of
Stimulus	Stimulus in a nervous system, a factor that triggers sensory transduction.
Critical period	Period of time during the life of an animal when imprinting can take place is a critical period.
Imprinting	Learning that is limited to a specific critical period in an animal's life and that is generally irreversible is called imprinting.
Endocrine system	The endocrine system is a set of internal organs involved in the secretion of hormones into the blood. These glands are known as ductless, which means they do not have tubes inside them.
Estrus	The estrus cycle refers to the recurring physiologic changes that are induced by reproductive hormones in most mammalian placental females (humans and great apes are the only mammals who undergo a menstrual cycle instead).
Testosterone	Testosterone is a steroid hormone from the androgen group. Testosterone is secreted in the testes of men and the ovaries of women. It is the principal male sex hormone and the "original" anabolic steroid. In both males and females, it plays key roles in health and well-being.
Territory	In ethology, sociobiology and behavioral ecology, the term territory refers to any geographical area that an animal of a particular species consistently defends against conspecifics (and, occasionally, animals of other species).
Brain	The part of the central nervous system involved in regulating and controlling body activity and interpreting information from the senses transmitted through the nervous system is referred to as the brain.
Gene	Gene refers to a discrete unit of hereditary information consisting of a specific nucleotide sequence in DNA . Most of the genes of a eukaryote are located in its chromosomal DNA; a few are carried by the DNA of mitochondria and chloroplasts.
Artificial selection	Artificial selection is the process of intentional or unintentional modification of a species through human actions which encourage the breeding of certain traits over others. When the process leads to undesirable outcome, it is called negative selection.
Fruit	A fruit is the ripened ovary—together with seeds—of a flowering plant. In many species, the fruit incorporates the ripened ovary and surrounding tissues.
Segregation	The separation of homologous chromosomes during mitosis and meiosis. Known as Mendel's theory of Segregation.
Bacterium	Most bacterium are microscopic and unicellular, with a relatively simple cell structure lacking a cell nucleus, and organelles such as mitochondria and chloroplasts. They are the most abundant of all organisms. They are ubiquitous in soil, water, and as symbionts of other organisms.
Larva	A free-living, sexually immature form in some animal life cycles that may differ from the adult in morphology, nutrition, and habitat is called larva.
Synapse	A junction, or relay point, between two neurons, or between a neuron and an effector cell. Electrical and chemical signals are relayed from one cell to another at a synapse.
Genetics	Genetics is the science of genes, heredity, and the variation of organisms.
Protein	A protein is a complex, high-molecular-weight organic compound that consists of amino acids

Go to **Cram101.com** for the Practice Tests for this Chapter.

joined by peptide bonds. They are essential to the structure and function of all living cells and viruses. Many are enzymes or subunits of enzymes.

Physiology	The study of the function of cells, tissues, and organs is referred to as physiology.
Pheromone	Chemical signal that works at a distance and alters the behavior of another member of the same species is called a pheromone.
Molecule	A molecule is the smallest particle of a pure chemical substance that still retains its chemical composition and properties.
Diffusion	Diffusion refers to the spontaneous movement of particles of any kind from where they are more concentrated to where they are less concentrated.
Invertebrate	Invertebrate is a term coined by Jean-Baptiste Lamarck to describe any animal without a spinal column. It therefore includes all animals except vertebrates (fish, reptiles, amphibians, birds and mammals).
Circadian rhythm	Circadian rhythm is the name given to the roughly 24 hour cycles shown by physiological processes in plants, animals, fungi and cyanobacteria.
Nerve	A nerve is an enclosed, cable-like bundle of nerve fibers or axons, which includes the glia that ensheath the axons in myelin.
Tissue	Group of similar cells which perform a common function is called tissue.
Vertebrate	Vertebrate is a subphylum of chordates, specifically, those with backbones or spinal columns. They started to evolve about 530 million years ago during the Cambrian explosion, which is part of the Cambrian period.
Pineal gland	The pineal gland is a small endocrine gland in the brain. It is located near the center of the brain, between the two hemispheres, tucked in a groove where the two rounded thalamic bodies join.
Gland	A gland is an organ in an animal's body that synthesizes a substance for release such as hormones, often into the bloodstream or into cavities inside the body or its outer surface.
Multicellular	Multicellular organisms are those organisms consisting of more than one cell, and having differentiated cells that perform specialized functions. Most life that can be seen with the naked eye is multicellular, as are all animals (i.e. members of the kingdom Animalia) and plants (i.e. members of the kingdom Plantae).
Translation	Translation is the second process of protein biosynthesis. In translation, messenger RNA is decoded to produce a specific polypeptide according to the rules specified by the genetic code.
Transcription	Transcription is the process through which a DNA sequence is enzymatically copied by an RNA polymerase to produce a complementary RNA. Or, in other words, the transfer of genetic information from DNA into RNA.
Cytoplasm	Cytoplasm refers to everything inside a cell between the plasma membrane and the nucleus; consists of a semifluid medium and organelles.
Migration	Migration occurs when living things move from one biome to another. In most cases organisms migrate to avoid local shortages of food, usually caused by winter. Animals may also migrate to a certain location to breed, as is the case with some fish.
Adaptation	A biological adaptation is an anatomical structure, physiological process or behavioral trait of an organism that has evolved over a period of time by the process of natural selection such that it increases the expected long-term reproductive success of the organism.

Predator	A predator is an animal or other organism that hunts and kills other organisms for food in an act called predation.
Species	Group of similarly constructed organisms capable of interbreeding and producing fertile offspring is a species.
Ecology	Ecology is the scientific study of the distribution and abundance of living organisms and how these properties are affected by interactions between the organisms and their environment.
Evolution	In biology, evolution is the process by which novel traits arise in populations and are passed on from generation to generation. Its action over large stretches of time explains the origin of new species and ultimately the vast diversity of the biological world.
Natural selection	Natural selection is the process by which biological individuals that are endowed with favorable or deleterious traits end up reproducing more or less than other individuals that do not possess such traits.
Testosterone	Testosterone is a steroid hormone from the androgen group. Testosterone is secreted in the testes of men and the ovaries of women. It is the principal male sex hormone and the "original" anabolic steroid. In both males and females, it plays key roles in health and well-being.
Territorial behavior	Behavior involved in establishing, defending, and maintaining a territory for food, mating, or other purposes is called territorial behavior.
Habitat	Habitat refers to a place where an organism lives; an environmental situation in which an organism lives.
Egg	An egg is the zygote, resulting from fertilization of the ovum. It nourishes and protects the embryo.
Larva	A free-living, sexually immature form in some animal life cycles that may differ from the adult in morphology, nutrition, and habitat is called larva.
Algae	The algae consist of several different groups of living organisms that capture light energy through photosynthesis, converting inorganic substances into simple sugars with the captured energy.
Sperm	Sperm refers to the male sex cell with three distinct parts at maturity: head, middle piece, and tail.
Sexual selection	Changes in males and females, often due to male competition and female selectivity leading to increased fitness are called sexual selection.
Insect	An arthropod that usually has three body segments , three pairs of legs, and one or two pairs of wings is called an insect. They are the largest and (on land) most widely-distributed taxon within the phylum Arthropoda. They comprise the most diverse group of animals on the earth, with around 925,000 species described
Reproduction	Biological reproduction is the biological process by which new individual organisms are produced. Reproduction is a fundamental feature of all known life; each individual organism exists as the result of reproduction by an antecedent.
Carnivore	An animal that eats a diet consisting solely of meat is referred to as a carnivore.
Population	Group of organisms of the same species occupying a certain area and sharing a common gene pool is referred to as population.
Inclusive fitness	Inclusive fitness encompasses conventional Darwinian fitness with the addition of behaviors that contribute to an organism's individual fitness through altruism. An organism's ultimate goal is to leave the maximum number of viable offspring possible, thereby keeping their genes

Go to **Cram101.com** for the Practice Tests for this Chapter.

present within a population.

Mole	The atomic weight of a substance, expressed in grams. One mole is defined as the mass of 6.02223×10^{23} atoms.
Predation	Interaction in which one organism uses another, called the prey, as a food source is referred to as predation.
Mammal	Homeothermic vertebrate characterized especially by the presence of hair and mammary glands is a mammal.
Territory	In ethology, sociobiology and behavioral ecology, the term territory refers to any geographical area that an animal of a particular species consistently defends against conspecifics (and, occasionally, animals of other species).
Protein	A protein is a complex, high-molecular-weight organic compound that consists of amino acids joined by peptide bonds. They are essential to the structure and function of all living cells and viruses. Many are enzymes or subunits of enzymes.
Diurnal	Diurnal is a term that refers to being active in the daytime.
Primate	A primate is any member of the biological group that contains all lemurs, monkeys, apes, and humans.

Desert	A desert is a landscape form or region that receives little precipitation - less than 250 mm (10 in) per year. It is a biome characterized by organisms adapted to sparse rainfall and rapid evaporation.
Species	Group of similarly constructed organisms capable of interbreeding and producing fertile offspring is a species.
Population	Group of organisms of the same species occupying a certain area and sharing a common gene pool is referred to as population.
Distribution	Distribution in pharmacology is a branch of pharmacokinetics describing reversible transfer of drug from one location to another within the body.
Reproduction	Biological reproduction is the biological process by which new individual organisms are produced. Reproduction is a fundamental feature of all known life; each individual organism exists as the result of reproduction by an antecedent.
Demography	Demography refers to the statistical study of population. It encompasses the study of the size, structure and distribution of populations, and how populations change over time due to births, deaths, migration and ageing.
Host	Host is an organism that harbors a parasite, mutual partner, or commensal partner; or a cell infected by a virus.
Habitat	Habitat refers to a place where an organism lives; an environmental situation in which an organism lives.
Population density	The number of organisms of a species per unit area is called population density.
Density-independent	Referring to any factor that limits a population's size and growth regardless of its density is density-independent.
Territorial behavior	Behavior involved in establishing, defending, and maintaining a territory for food, mating, or other purposes is called territorial behavior.
Carrying capacity	Carrying capacity is the measure of habitat to indefinitely sustain a population at a particular density.
Spore	Spore refers to a differentiated, specialized form that can be used for dissemination, for survival of adverse conditions because of its heat and dessication resistance, and/or for reproduction. They are usually unicellular and may develop into vegetative organisms or gametes. A spore may be produced asexually or sexually and are of many types.
Evolution	In biology, evolution is the process by which novel traits arise in populations and are passed on from generation to generation. Its action over large stretches of time explains the origin of new species and ultimately the vast diversity of the biological world.
Natural selection	Natural selection is the process by which biological individuals that are endowed with favorable or deleterious traits end up reproducing more or less than other individuals that do not possess such traits.
Extinction	In biology and ecology, extinction is the ceasing of existence of a species or group of taxa. The moment of extinction is generally considered to be the death of the last individual of that species. The death of all members of a species is extinction.

Species	Group of similarly constructed organisms capable of interbreeding and producing fertile offspring is a species.
Predator	A predator is an animal or other organism that hunts and kills other organisms for food in an act called predation.
Commensalism	Symbiotic relationship in which one species is benefited and the other is neither harmed nor benefited is commensalism.
Intertidal zone	The part of a beach that lies between average high tide and average low tide is called the intertidal zone.
Extinction	In biology and ecology, extinction is the ceasing of existence of a species or group of taxa. The moment of extinction is generally considered to be the death of the last individual of that species.The death of all members of a species is extinction.
Root	In vascular plants, the root is that organ of a plant body that typically lies below the surface of the soil. However, this is not always the case, since a root can also be aerial (that is, growing above the ground) or aerating (that is, growing up above the ground or especially above water).
Host	Host is an organism that harbors a parasite, mutual partner, or commensal partner; or a cell infected by a virus.
Bacteria	The domain that contains procaryotic cells with primarily diacyl glycerol diesters in their membranes and with bacterial rRNA. Bacteria also is a general term for organisms that are composed of procaryotic cells and are not multicellular.
Population	Group of organisms of the same species occupying a certain area and sharing a common gene pool is referred to as population.
Virus	Obligate intracellular parasite of living cells consisting of an outer capsid and an inner core of nucleic acid is referred to as virus. The term virus usually refers to those particles that infect eukaryotes whilst the term bacteriophage or phage is used to describe those infecting prokaryotes.
Batesian mimicry	A type of mimicry in which a species that a predator can eat looks like a different species that is poisonous or otherwise harmful to the predator is called Batesian mimicry.
Herbivore	A herbivore is an animal that is adapted to eat primarily plant matter
Mullerian mimicry	A mutual mimicry by two species, both of which are poisonous or otherwise harmful to a predator is a Mullerian mimicry.
Insect	An arthropod that usually has three body segments , three pairs of legs, and one or two pairs of wings is called an insect. They are the largest and (on land) most widely-distributed taxon within the phylum Arthropoda. They comprise the most diverse group of animals on the earth, with around 925,000 species described
Microorganism	A microorganism is an organism that is so small that it is microscopic (invisible to the naked eye). They are often illustrated using single-celled, or unicellular organisms; however, some unicellular protists are visible to the naked eye, and some multicellular species are microscopic.
Cellulose	A large polysaccharide composed of many glucose monomers linked into cable-like fibrils that provide structural support in plant cell walls is referred to as cellulose.
Angiosperm	Flowering plant that produces seeds within an ovary that develops into a fruit is referred to as an angiosperm.
Pollination	In seed plants, the delivery of pollen to the vicinity of the egg-producing megagametophyte

is pollination.

Sugar	A sugar is the simplest molecule that can be identified as a carbohydrate. These include monosaccharides and disaccharides, trisaccharides and the oligosaccharides. The term "glyco-" indicates the presence of a sugar in an otherwise non-carbohydrate substance.
Mutualism	Mutualism is an interaction between two species in which both species derive benefit.
Color vision	Ability to detect the color of an object, dependent on three kinds of cone cells is called color vision.
Coevolution	Coevolution is the mutual evolutionary influence between two species that become dependent on each other.
Flower	A flower is the reproductive structure of a flowering plant. The flower structure contains the plant's reproductive organs, and its function is to produce seeds through sexual reproduction.
Ovary	In the flowering plants, an ovary is a part of the female reproductive organ of the flower or gynoecium.
Pollen	The male gametophyte in gymnosperms and angiosperms is referred to as pollen.
Keystone species	Species that are not usually abundant in a community yet exert strong control on community structure by the nature of their ecological roles or niches are keystone species.
Climate	Weather condition of an area including especially prevailing temperature and average daily/yearly rainfall over a long period of time is called climate.
Prairie	Prairie refers to a biome, located in the centers of continents, that supports grasses; also called grassland.
Urine	Concentrated filtrate produced by the kidneys and excreted via the bladder is called urine.
Leaf	In botany, a leaf is an above-ground plant organ specialized for photosynthesis. For this purpose, a leaf is typically flat (laminar) and thin, to expose the chloroplast containing cells (chlorenchyma tissue) to light over a broad area, and to allow light to penetrate fully into the tissues.
Predation	Interaction in which one organism uses another, called the prey, as a food source is referred to as predation.
Ecological succession	Ecological succession refers to the process of biological community change resulting from disturbance; transition in the species composition of a biological community, often following a flood, fire, or volcanic eruption.
Nitrogen fixation	Nitrogen fixation is the process by which nitrogen is taken from its relatively inert molecular form (N_2) in the atmosphere and converted into nitrogen compounds useful for other chemical processes.
Nitrogen	Nitrogen is a chemical element in the periodic table which has the symbol N and atomic number 7. Commonly a colorless, odorless, tasteless and mostly inert diatomic non-metal gas, nitrogen constitutes 78.08% percent of Earth's atmosphere and is a constituent of all living tissues. Nitrogen forms many important compounds such as amino acids, ammonia, nitric acid, and cyanides.
Fixation	Fixation in population genetics occurs when the frequency of a gene reaches 1. Fixation in biochemistry, histology, cell biology and pathology refers to the technique of preserving a specimen for microscopic study, making it intact and stable, but dead.

Go to **Cram101.com** for the Practice Tests for this Chapter.

379

Trade winds	The trade winds are a pattern of wind found in bands around the Earth's equatorial region. The trade winds are the prevailing winds in the tropics, blowing from the high-pressure area in the horse latitudes towards the low-pressure area around the equator.
Climate	Weather condition of an area including especially prevailing temperature and average daily/yearly rainfall over a long period of time is called climate.
Specific heat	The amount of energy required to raise the temperature of 1 gram of a substance by 1 °C is specific heat.
Photosynthesis	Photosynthesis is a biochemical process in which plants, algae, and some bacteria harness the energy of light to produce food. Ultimately, nearly all living things depend on energy produced from photosynthesis for their nourishment, making it vital to life on Earth.
Trophic level	In ecology, the trophic level is the position that an organism occupies in a food chain - what it eats, and what eats it.
Herbivore	A herbivore is an animal that is adapted to eat primarily plant matter
Food web	A network of interconnecting food chains is referred to as a food web.
Biomass	Biomass is organic non-fossil material, collectively. In other words, biomass comprises the mass of all biological organisms, dead or alive, excluding biological mass that has been transformed by geological processes into substances such as coal or petroleum.
Ecosystem	In general terms an ecosystem can be thought of as an assemblage of organisms (plant, animal and other living organisms living together with their environment, functioning as a loose unit. That is, a dynamic and complex whole, interacting as an "ecological unit".
Detritus	Detritus is organic waste material from decomposing dead plants or animals.
Decomposition	Decomposition refers to the reduction of the body of a formerly living organism into simpler forms of matter.
Species	Group of similarly constructed organisms capable of interbreeding and producing fertile offspring is a species.
Crop	An organ, found in both earthworms and birds, in which ingested food is temporarily stored before being passed to the gizzard, where it is pulverized is the crop.
Pollution	Any environmental change that adversely affects the lives and health of living things is referred to as pollution.
Transformation	Transformation is the genetic alteration of a cell resulting from the introduction, uptake and expression of foreign genetic material (DNA or RNA).
Element	A chemical element, often called simply element, is a chemical substance that cannot be divided or changed into other chemical substances by any ordinary chemical technique. An element is a class of substances that contain the same number of protons in all its atoms.
Respiration	Respiration is the process by which an organism obtains energy by reacting oxygen with glucose to give water, carbon dioxide and ATP (energy). Respiration takes place on a cellular level in the mitochondria of the cells and provide the cells with energy.
Upwelling	Upward movement of deep, nutrient-rich water along coasts is referred to as upwelling.
Wavelength	The distance between crests of adjacent waves, such as those of the electromagnetic spectrum is wavelength.
Radiation	The emission of electromagnetic waves by all objects warmer than absolute zero is referred to as radiation.
Ion	Ion refers to an atom or molecule that has gained or lost one or more electrons, thus

acquiring an electrical charge.

Bacteria	The domain that contains procaryotic cells with primarily diacyl glycerol diesters in their membranes and with bacterial rRNA. Bacteria also is a general term for organisms that are composed of procaryotic cells and are not multicellular.
Fossil	A preserved remnant or impression of an organism that lived in the past is referred to as fossil.
Global warming	A slow but steady rise in Earth's surface temperature, caused by increasing concentrations of greenhouse gases in the atmosphere is referred to as global warming.
Cellulose	A large polysaccharide composed of many glucose monomers linked into cable-like fibrils that provide structural support in plant cell walls is referred to as cellulose.
Ammonia	Ammonia is a compound of nitrogen and hydrogen with the formula NH_3. At standard temperature and pressure ammonia is a gas. It is toxic and corrosive to some materials, and has a characteristic pungent odor.
Algae	The algae consist of several different groups of living organisms that capture light energy through photosynthesis, converting inorganic substances into simple sugars with the captured energy.
Acid	An acid is a water-soluble, sour-tasting chemical compound that when dissolved in water, gives a solution with a pH of less than 7.
Nitrifying bacteria	Several kinds of bacteria capable of converting ammonia to nitrite, or nitrite to nitrate is referred to as nitrifying bacteria.
Nitrogen cycle	Nitrogen cycle refers to continuous process by which nitrogen circulates in the air, soil, water, and organisms of the biosphere.
Nitrogen	Nitrogen is a chemical element in the periodic table which has the symbol N and atomic number 7. Commonly a colorless, odorless, tasteless and mostly inert diatomic non-metal gas, nitrogen constitutes 78.08% percent of Earth's atmosphere and is a constituent of all living tissues. Nitrogen forms many important compounds such as amino acids, ammonia, nitric acid, and cyanides.
Filtration	Filtration involved in passive transport is the movement of water and solute molecules across the cell membrane due to hydrostatic pressure by the cardiovascular system.
Bacterium	Most bacterium are microscopic and unicellular, with a relatively simple cell structure lacking a cell nucleus, and organelles such as mitochondria and chloroplasts. They are the most abundant of all organisms. They are ubiquitous in soil, water, and as symbionts of other organisms.

Species	Group of similarly constructed organisms capable of interbreeding and producing fertile offspring is a species.
Vascular	In botany vascular refers to tissues that contain vessels for transporting liquids. In anatomy and physiology, vascular means related to blood vessels, which are part of the Circulatory system.
Continental drift	Continental drift refers to movement of continents with respect to one another over the earth's surface.
Evolution	In biology, evolution is the process by which novel traits arise in populations and are passed on from generation to generation. Its action over large stretches of time explains the origin of new species and ultimately the vast diversity of the biological world.
Transformation	Transformation is the genetic alteration of a cell resulting from the introduction, uptake and expression of foreign genetic material (DNA or RNA).
Molecular clock	The molecular clock is a technique in genetics, which researchers use to date when two species diverged. It deduces elapsed time from the number of minor differences between their DNA sequences.
Taxon	A taxon, is a grouping of organisms (named or unnamed). Once named, a taxon will usually have a rank and can be placed at a particular level in a hierarchy.
Fossil	A preserved remnant or impression of an organism that lived in the past is referred to as fossil.
Extinction	In biology and ecology, extinction is the ceasing of existence of a species or group of taxa. The moment of extinction is generally considered to be the death of the last individual of that species. The death of all members of a species is extinction.
Evergreen	In botany, an evergreen plant is a plant that retains its leaves all year round, with each leaf persisting for more than 12 months. This contrasts with deciduous plants, which completely lose all their foliage for part of the year, becoming bare and leafless.
Biome	A biome is a major regional group of distinctive plant and animal communities best adapted to the region's physical environment. The concept of a biome emphasizes the cohesion or correlation among species groups, soils, and climate, rather than any one of them singly.
Climate	Weather condition of an area including especially prevailing temperature and average daily/yearly rainfall over a long period of time is called climate.
Tundra	A biome at the northernmost limits of plant growth and at high altitudes, characterized by dwarf woody shrubs, grasses, mosses, and lichens is referred to as tundra.
Permafrost	Permanently frozen ground usually occurring in the tundra, a biome of Arctic regions is referred to as permafrost.
Cone	Cone refers to a reproductive structure of gymnosperms that produces pollen in males or eggs in females.
Temperate deciduous forest	A type of forest located throughout midlatitude regions where there is sufficient moisture to support the growth of large, broadleaf deciduous trees is referred to as temperate deciduous forest.
Deciduous	A term used to describe trees that lose their leaves at the end of the growing season is a deciduous.
Temperate	In virology, temperate refers to the life cycle a phage is able to perform. The phage usually integrates its genome into its host chromosome, becoming a prophage.
Desert	A desert is a landscape form or region that receives little precipitation - less than 250 mm

(10 in) per year. It is a biome characterized by organisms adapted to sparse rainfall and rapid evaporation.

Chaparral	Chaparral refers to a biome dominated by spiny evergreen shrubs adapted to periodic drought and fires; found where cold ocean currents circulate offshore, creating mild, rainy winters and long, hot, dry summers.
Photosynthesis	Photosynthesis is a biochemical process in which plants, algae, and some bacteria harness the energy of light to produce food. Ultimately, nearly all living things depend on energy produced from photosynthesis for their nourishment, making it vital to life on Earth.
Genus	In biology, a genus is a taxonomic grouping. That is, in the classification of living organisms, a genus is considered to be distinct from other such genera. A genus has one or more species: if it has more than one species these are likely to be morphologically more similar than species belonging to different genera.
Savanna	A biome dominated by grasses and scattered trees is called savanna. A savanna is sometimes a transitional zone, occurring between forest or woodland regions and grassland regions.
Thorn	Thorn refers to a hard, pointed outgrowth of a stem; normally a modified branch.
Tropical deciduous forest	Tropical deciduous forest refers to a biome with pronounced wet and dry seasons and plants that must shed their leaves during the dry season to minimize water loss.
Salt	Salt is a term used for ionic compounds composed of positively charged cations and negatively charged anions, so that the product is neutral and without a net charge.
Littoral	A littoral is the region near the shoreline of a body of fresh or salt water.
Algae	The algae consist of several different groups of living organisms that capture light energy through photosynthesis, converting inorganic substances into simple sugars with the captured energy.
Mammal	Homeothermic vertebrate characterized especially by the presence of hair and mammary glands is a mammal.

Extinction	In biology and ecology, extinction is the ceasing of existence of a species or group of taxa. The moment of extinction is generally considered to be the death of the last individual of that species. The death of all members of a species is extinction.
Species	Group of similarly constructed organisms capable of interbreeding and producing fertile offspring is a species.
Population	Group of organisms of the same species occupying a certain area and sharing a common gene pool is referred to as population.
Ecosystem	In general terms an ecosystem can be thought of as an assemblage of organisms (plant, animal and other living organisms living together with their environment, functioning as a loose unit. That is, a dynamic and complex whole, interacting as an "ecological unit".
Habitat	Habitat refers to a place where an organism lives; an environmental situation in which an organism lives.
Vascular	In botany vascular refers to tissues that contain vessels for transporting liquids. In anatomy and physiology, vascular means related to blood vessels, which are part of the Circulatory system.
Fungus	A fungus is a eukaryotic organism that digests its food externally and absorbs the nutrient molecules into its cells.
Crop	An organ, found in both earthworms and birds, in which ingested food is temporarily stored before being passed to the gizzard, where it is pulverized is the crop.
Climate	Weather condition of an area including especially prevailing temperature and average daily/yearly rainfall over a long period of time is called climate.
Global warming	A slow but steady rise in Earth's surface temperature, caused by increasing concentrations of greenhouse gases in the atmosphere is referred to as global warming.
Endangered species	An endangered species is a population of organisms (frequently but not always a taxonomic species) which is either (a) so few in number or (b) threatened by changing environmental or predation parameters that it is at risk of becoming extinct.
Endemic species	In biology and ecology endemic species means exclusively native to a place or biota, in contrast to cosmopolitan or introduced.
Biodiversity	Biodiversity is the totality of genes, species, and ecosystems of a region.
Deciduous	A term used to describe trees that lose their leaves at the end of the growing season is a deciduous.
Wetland	A wetland is an environment "at the interface between truly terrestrial ecosystems...and truly aquatic systems...making them different from each yet highly dependent on both."
Cancer	Cancer is a class of diseases or disorders characterized by uncontrolled division of cells and the ability of these cells to invade other tissues, either by direct growth into adjacent tissue through invasion or by implantation into distant sites by metastasis.

Go to **Cram101.com** for the Practice Tests for this Chapter.

389

Printed in the United Kingdom
by Lightning Source UK Ltd.
114021UKS00001B/5